Calligraphy, Hazel Beney

'To marvel at nature is what I am here for,' says Goethe. And, certainly, the continual wonder and beauty of the world around can gild one's thoughts and enrich one's living.

Here, is offered a bouquet 'of other men's flowers', with my own blossoms to 'bind them together'.

It is hoped that this appreciation, however simple and unpretentious, will revive, in 'the winters of our discontent', something of the joy and delight that unceasingly surrounds us, in this England of ours.

B.N.H.

287

England

MERLIN BOOKS LTD.
Braunton   Devon

To
H. and P.

ISBN 0 86303 202-8
Printed in England by A. Wheaton & Co.Ltd., Exeter, Devon

JANUARY

'All seasons shall be dear to thee.'
'And beauty's self she is
When all her robes are gone.'

What is a New Year if not a challenge; an unsullied beginning to an exciting journey; an offering of time, enhanced by wildlife and its attendant wonders. And, what better advice to follow than Pope's when he adjures us to 'First, follow Nature.'

From Janus, god of opening, of beginning, January is the gateway to the year. Ushered in by the insistent 'music's laughter' of bells, we stand with Nature on the threshold of Spring. Though superficially dark and secretive, this month, the first, is like an open window letting in light, little by little, on the pageant of fresh, new life. Drear morning skies of January often give way to flaming beauty of sunset. Certainly, it is the season when, with C.D. Lewis, we can believe 'that we hear again and firmly walk again, in these our Winter days.'

Fittingly enough, January of 1981 begins for me with the sunrise. Wakened by the strange, awesome call of a dog fox, 'howling from his chosen lair', I rise at 6 o'clock to be richly rewarded. Skeins of mist lightly cloud the fields to hand, tangled, in the tree-tops. Slowly, the sky above them lightens, the veil lifts and little peach clouds show in the dawn blue. Over Merry Hill, sunrise, and a delicate glow of pearl and palest primrose suffuses the sleeping world.

The second day brings a splashing New Year. Heavy rains and rivers burst their banks, with the Severn and the Avon abnormally high and Christchurch Abbey marooned in a sea of water. Roads are blocked, fens and fields are flooded. The first Sunday in the New Year, a day which was once known as Plough Sunday, when ploughs were brought into church and blessed. The following day was a time to bedeck oneself and hold festival.

Despite the weather near by, January brings its own pageantry, its partly hidden treasures, more precious for their rarity and transcience. I find a red rose, frail and fragile, but still a rose; a berry, withering but bright; a wren, hungry, but still singing; a speed-well, blue; a slender catkin, bravely swinging and growing golden. January may be a resting time, but it is still a breathing time.

'Not from yesterday or the day before, do
these things come, but from everlasting;
and no one knows from whence they took
their beginning.'

Black ice with its dangers follows the flooding but, after a day or two, normal wintry conditions prevail. More treasures reveal themselves in my garden; the yellow of Winter jasmin, a sprig of purple heather and

my robin sings. Despite Gilbert White's condemnation, saying, 'He does much mischief in the garden to the Summer fruit,' the robin is always welcome.

For a few days the air remains bitingly cold and fresh. At first glance, Merry Hill seems empty and desolate. My garden appears forsaken, forlorn. The only movement from the Pools is among the quivering reeds. But, at last, the coppice can be seen from end to end and the still air brings the plaintive cry of a ewe and the bleat of her lamb. The wide fields wear their Winter green and every tree is unadorned. Lifting bare 'bones' to the frosty sky, each tree has its own filigree, patterned in graceful black twigs.

Across the meadow, in green pools of light, wide boughs cast clean-cut shadows, as pure as reflections. The tranquillity of Winter lies on the slopes and the silent landscape stands quiet and starkly beautiful against a cold sky.

Above my head, the magpies' nests are sited, large and untidy and high, and beyond is a sudden conclave of these birds. There is much clamorous flight and trailing of long tails, in a follow-my-leader ritual. The sun brings gulls and peewits to feed in the fields. Black and ragged rooks join them. A sudden disturbance and they all rise in the air, the rooks being much swifter in movement than the rest.

Deep in the hedgerow, a shrew hides, after tunnelling for worms and insects and caterpillars, and lies cautiously dormant. Moss and lichen provide cover for tiny things and in the wall's crevices insects and wasps are hiding. Across the field, I sight a hare racing for shelter. Like the squirrel, he has grown white patches of fur for camouflage.

At the foot of the bare hawthorn, I find wild carrot just beginning to grow, while a few chickweed and groundsel plants show minute buds. The biggest buds are on the horse-chestnut.

Late afternoon, and trees etched starkly against a sky of dove grey. A sudden darkening and a fox, under the hedge, slinks and crawls on his stomach, seeking food. With his gold-red body, fine head and splendid tail, he is a magnificent sight. Soon, his blind babies will be born. From time immemorial the fox has persisted, acutely aware, against all odds, eating rats, mice, birds, rabbits, snakes, frogs and hedgehogs. Now, in Winter's deep frosts, he has to hunt for beetles and snails and is even found in the heart of London, in litter-boxes and dustbins. By some, he is believed to be less destructive than a carrion crow, only eating chickens when fed to him as a cub. He is a master of trickery and, like the cat, prowls by night. Strangely, there have been reports of a friendship between a fox and a cat.

Frost has now burnished the vivid tints of Autumn, but a few reddish leaves still cling to the brambles and the last large berries to the deadly nightshade. Low in the hedges of the field, blackened feathers of fleece wave. Withered seed-pods are under my feet and bronzed leaves drift before me as I go homeward.

The first week ends with heavy grey skies and, almost unnoticed white flakes fall, hesitantly but thickly. And, soon, 'the world is wondrous with fallen snow.' A rising wind and I am enclosed on the Hill in a blinding, whirling mass, which quickly wraps its white mantle over everything, obliterating contours, blanketing sound. Before me, boughs seem suddenly to rise in an etching of blacks, browns, purples; hedgerows twist in a rusting haze and the depths of the coppice are wine-dark.

Driven by the wind in level clouds, every snowflake, in its perfection brings more beauty and I am lost in an intimate world of wonder, believing with Bewick that 'to be in a whirlwind of snow, while the tempest howls above my head, is sublimity itself.'

The following day, and the blanket deepens. Softly, the snowflakes still fall,

'On winds of space, like flowers to blow
In a wilderness of blue.'

A perfect Winter's day on the twelfth, with sunshine sparkling on the icy snow and trees of the coppice, casting long, lavender shadows over the hidden lawn. Shrubs of the garden are weighted down; a rhododendron is uprooted and fir boughs sweep the earth under their burden. Remnants of cabbages are frosted and crisply curled. In some parts, acres of them are buried and rotting under ice and flood water.

Branches in the coppice are patterned in ebony, with powdery crystals of blue and silver. Sudden dazzling rays from the sun and, magically, the trees are full of birds, hungry and watchful. Blue tits come daintily for nuts; wrens cry and a solitary thrush, with the wind ruffling his feathers. My blackbird appears, looking twice as big with his fluffed out wings. The berries he loves have almost gone and he has to look for spiders and grubs. A magpie, with his awkward, heavy gait, drops among them on the snowy lawn and they all retreat, hurriedly.

By the afternoon, the purity of the snow is yellowing under the strange, flat light of a heatless sun and, in a world of seeming weightlessness, as though reluctant to lose their new-found glory, boughs let fall their burden of snow. A brief visit from a jay, looking magnificent against the snow, with his blue and black barred wings and his striking crest. He does not linger, since most of the birds' food has gone.

Four days of bitter weather, yet my garden wears a new beauty in the accentuated and strange grace of the sumach tree, where one long, orange leaf still clings.

Called the Tree of Heaven, it is reminiscent of a Japanese picture. A few blades of grass appear through the snow, stiffened with white frost. From an azalea bush my wren sings.

'So wild and shrill she cries
To fill the skies.'

The golden cypress, still in snowy layers, tapers upwards and the turquoise has deepened on the silvery fir. Berberis berries are greatly depleted by the hungry birds, but delicate, slim buds show already on the birch and branches of the cherry tree are polished and shining.

Another welcome discovery. My daphne has clusters of purple-pink flowers, not yet fully open, and the green of Winter hellebores seems to glow, against decaying leaf mould. As I watch, a family of blue tits come to enjoy their bag of nuts. These pretty birds are mostly an azure blue, but their black and white faces have earned them the name of 'nun'. A sharp-eyed squirrel watches, too, but not to admire. From the coppice he suddenly pounces and outdoes the scattered blue tits in manoeuvrability, swinging and tight-rope walking along the hanging string to get at the bag of nuts.

When he has gone the blue tits return, to enjoy a shower in the melting snow of a stone basin, their fluttering wings sending sparkling sprays of water into the air.

After two days the snows have vanished and, satisfied with a feast on the berries of cotoneaster, my blackbird sings as sweetly as though it were already Spring. Near him, I find a polyanthus, dashed but alive and a spray of pink heather. Warm and welcoming is the yellow of Winter jasmine and wych-hazel. And, from crinkled, dead leaves, a small red rose glows.

Apart from a few tiny buds, the wide-flung arms of the beech in the coppice are bare, but not the floor at its feet. In layers of golden brown it is thick with fallen foliage and leaf mould.

Several rooks, from their nests among the tiny purple buds of the ash, have flown here from Merry Hill. They dearly love the earthworms of the lawn.

These Winter birds are curious creatures. Like humans, they try not to miss anything. They do not go far, but are content with short, low flights under bushes. When they feed on the lawn, the perky sparrows quickly gather round to investigate and a lone thrush or robin also joins them, a little apart. The sparrows are quarrelsome, but not noisy like the starlings, who appear in an untidy crowd and waddle round. The farmer likes these birds since they clear his ground of leather-jackets and grubs. They can drop on an adder from the air and kill it instantly with their sharp beaks. And yet, even starlings are being tamed to eat from the hand, one taking cheese from the bird-man of Buckingham Palace.

When there is a disturbance, however, I find that

Some birds are great mimics. Not only do parrots and mynahs imitate sounds, but also starlings and thrushes, while more than one jackdaw has had its tongue cut, in the past, in a misguided attempt to make it talk. My blackbird repeats a whistle, while the stonechat has other sounds to his repertoire, apart from that of falling stones.

The last week, and the intense cold is lifting slightly. The high moors of Yorkshire stretch before me in a vast flow of open, treeless space. Far away, the sky is as 'the serene of heaven', but over the near hills, cloud and shadow chase each other in a swiftly changing scene. The heights are lonely, desolate. It is here, among the ancient rocks that have been carved by the wind and rain of centuries, that wild things live their lives in wild places. From every side comes the sound of water, rippling and sparkling from the snows of Winter in deep shafts and narrow pools. The landscape is empty, having been cleared long ago by the burning and felling of sparse trees.

The river, swollen with the snows and heavy rainfall, flows darkly still, carrying on its surface drifting leaves and in its waters, grayling, roach, eel, and trout. On its banks live sandmartins and oyster-catchers, visible today because of hunger. In the deep shallows a curlew flies down to wade and a dipper makes his curious motions. Both are being watched from above by a buzzard on the wing. Milkwort and knapweed have blossomed and died and numerous lady-smock will later flower, with their attendant orange-tip butterflies, which are common in these parts.

Near a limestone bed on the river, with its huge, down-flung boulders, a badger has his set. Through binoculars, his black and white face and dark grey fur show plainly. Withered tormentil grows in the grass of the verge.

Until recently, otter hunts were followed on foot, accompanied by beagles, but these are now banned. Foxes, which are particularly wild here, are hunted from the 'scar' woods of the Dale, where holly and mountain ash still show berries and the guelder rose later grows.

Wide fields rise from the river, enclosed by ancient stone walls, wonderfully wrought. Many brown hares hunt in the lower pastures and some wild mink, which are a nuisance to farmers. Higher, on a rock, a short-eared owl sits immobile, searching, keen-eyed for food.

Across the slopes of the level hills, among patches of heather, a few sheep graze, huddling together for warmth. When the earlier snows were deep, carrion crows pecked at the snow-bound flock. A flash of green, red and chestnut. It is the iridescent plumage of a cock pheasant, soon lost among the heather.

Through deep and secret valleys and mist-filled hollows, I climb, where becks, bracken-rimmed, worm their way between rocks to tumble down miniature headlands and small ravines. The whole landscape is threaded with thin and silver ribbons of water.

Marshy spots among the boulders show the vivid

starlings are cowardly birds, being the first to scatter, while small birds creep quietly in to feed. The lawn is cleared entirely when a magpie descends, raucous, to stalk across the lawn, with glossy plumage and long tail trailing. I watch as he returns to his high nest in the coppice to clear the thick snow from the top of his slovenly home, in preparation for a family. My black-bird replaces him. One of the gifts of Winter is the sheen of the sun on his throat and the russet-breasted beauty of his wife.

My robin, in his own good time, perches on his low wall, when he sees me at the window and waits trustingly. The severe weather has brought gulls inland, foraging. It is reported that some hungry ones have dive-bombed people near their homes.

Rare visitors to my garden are the greenfinch, great tit and coal tit, who loves fat. The great spotted wood-pecker, after eating seeds from the Scotch pine, occasionally drops, in his dipping flight to enjoy fat and wholemeal bread, but he is a recent visitor to my garden.

Undoubtedly, just as some animals are people lovers, so birds seem to need companionship. My blackbird and robin regularly follow me about the garden and, on rare occasions, inside the door.

A neighbour going on holiday made this request, "Their food is being taken care of, but would you please go in each day to sing to them?" I did my best to oblige, though the birds were a caged budgerigar and a cockatoo.

Cry of the Curlew

Where curlews cry, and kine
Creep heavy to the swollen stream,
And cornfields climb in widening wealth
To the crowned hill above,
Where pastures mount & pine-trees lift,
  Oh, there do curlews cry!

And there they swoop and sweep & grieve
With songs of mournful sadness,
In liquid notes that rise and lift,
Dissolve and fade, lone and forlorn,
O'er moorland wastes and
            deep streams tumbling
  Oh, there do curlews cry!

And, where the mild sheep gathering go
With shepherd, patient, solitary, slow,
    Why, there do curlews cry!
And, in the mists of morning's waking,
Through sunbeams,
       through deep shadows falling,
The lonely world they fill with calling,
  Oh, there do curlews cry!

Winter

Ice and fog and
Stormy weather:
Stars of jasmine
Sprays of heather.

Blackbird's bill
And robin's red:
Blue tits waiting
To be fed:

Curlew's call and
Cry of owl,
Wonder of the
    waterfowl:

Blackthorn bloom-
-ing on the hill:
Little lambs that
Bleat and shrill:

Berries, catkins,
Whirl of snow:
Flight of geese,
And sunset's glow.

colour of mosses and the withered stems of horseshoe vetch which spreads over the rocks. Scant moorgrass and silver hair-grass grow from the little swamps and round the rushy bogs, a thread of bright water runs, clear and sparkling.

Over on the barren heights snow lingers beneath white clouds which move swiftly, throwing long shadows. The sun, faintly warm, is beginning to crack the ice and slowly melt the snow.

From the brow of the hill a strange, rasping cry comes. It is a grouse, that native of the moors, whose sound is so evocative of these upland plains. As he leaves his eyrie in a solitary, stunted pine, a buzzard makes a mewing call. He has used plentiful sheeps' wool to line his nest. Now he is poised high in flight, watching for baby rabbits and any snakes that have ventured from their hidden hibernation. Once, these grouse were the size of a turkey.

Sadly, the red kite has left these moors and the golden eagle is rare. This eagle lives near the otter and often finishes the depleted remains of a fish that the otter has left. Hovering over the heather, I sight a merlin, our smallest falcon, which still haunts these parts, although some have left on migration.

Twilight comes early on the moors and the sky grows violet and dusky. Like sleeping ghosts, the lone hills crouch in the rising mists and the first star appears, lonely in immensity. A slight stirring, almost a presence, disturbs the air, rustling the heather and the cold grass. Clouds of ominous grey flee across the heavens like an out-going tide.

More barren than ever now and suddenly bitterly cold, the moors are a scene of desolation. Solitude dwells in the startlingly glistening outcrops of rock, in the mounting hills and the skies that touch them, leaning towards them as though sharing in the bare and utter loneliness. And, through the silence, comes the poignant, heart-searching cry of the curlew.

Home again and the month draws to a close. Early morning on Merry Hill is cold, clear and frost-rimmed. Frozen like snow, the ground is white with hoar frost and the furrows hard to the feet. Across to Oxhey and Harrow, the sloping fields show brown where the stout grass breaks through the rime. On the Hill's brow, far hedges and trees are hazy and dark.

The red breast of a robin stands startlingly bright against the frost of the five-barred gate. Where hedge parsley still stands withered in the shrivelled mould of the ditch, there is a little movement and a titmouse, long-tailed and hungry, peeps out, but quickly scuttles to shelter. A few last, brown leaves blow about before me and from above comes the loud cawing and noisy flight of rooks. Since the line of lofty elms, that for many years has crowned the ridge, patterning its silhouette against the clouds, has almost gone, these birds are losing their home. The farmers, however, will not be sorry to see the rooks go, as they do so much damage to young Spring crops.

Standing, gaunt and bare, the skeleton elms, leafless forever, wait for the final onslaught of the voracious grubs that devour them. It is sad to say good-bye to another familiar and crested landmark, gone with the ancient and splendid elm that shaded the village pond and added grace to the old church. Happily, in his picture, 'Our Village' this beloved elm has been perpetuated by the artist, Herkomer who lived here.

My wren in the garden does not forsake me, singing in rain, frost and mist. As she creeps about in dense bushes for insect larvae, her wings make a little whirring noise. In the bitterest weather, she nestles close to her large family for warmth.

January is ending. On my way to the Pools and a frosty day. As I leave, an early blackbird flies from his roost and perches on the chimney top, trilling his song, quite unaware that it is Winter and heedless of the cold.

My little avenue of limes, as I pass through, looks dank and dreary, but under the bare trees, I discover a yellow-budded dandelion and a closed daisy head, half buried in debris. A coltsfoot, too, which stores its food in an underground system, is almost in flower. Numerous shining buds are forming on the willow which, free of the snows, droops gracefully as though on a willow pattern plate.

The air of the Pools is cold, ice-captured. Frost covers everything. Though austere, it is a beautiful scene, suggestive of a Japanese print, such as Shelley must have seen, when he exclaimed, 'I love all forms of the radiant frost.'

A glinting sun throws a shimmering light on the water's edge, silver boles of birch lean low to throw blue and mauve shadows, curving clear-cut to meet their own icy reflections. Like miniature trees, each blade and plant shines in frosted perfection and fossil-like designs gleam on the ice of the Pool, in silver and blue.

The otter is hidden in his holt. The heron and the kingfisher, who cannot pierce this thick ice for food, have mostly moved away southwards though one has stayed. Only the ducks show movement in this world

of white, vainly seeking shallow waters for their up-ending and managing to look, at the same time, both ludicrous and lovely. But the pintail duck and the glossy mallard have to prepare for their babies, which are born this month or in early February.

Among the watercress, dead and stunted, the brave little wagtail has returned to Winter here. Like white icing the grass verge sparkles, twigs are etched in diamonds and ebony and small birds wistfully haunt yesterday's waters and peck disconsolately at their icy seal.

On this last of the month I stand as though waiting. The Pools are quiet, listening, while across the silence of the ice comes the sweet song of a blackbird, telling of the Spring and quite forgetful that it is still January.

'The blackbird plays but a boxwood flute, but I love him the best of all.'

FEBRUARY

'On some still day, when the year hangs between Winter and Spring and heaven is full of light.'

January has melted into February and her pageantry is decidedly muted. But, faintly, the year's beauty begins to emerge. There is a stirring of much hidden life. Freed, in part, from the grip of Winter, rivers and streams now are fast running, clearer and deeper. This movement of water over boulders, round bends, among reeds; this hissing and swishing from the floods of melting snows and rainfall, is synonymous with Spring. Insect and bird, bird and beast, beast and man, in step with the seasons, all share in the unceasing sequence of life in Nature. And,

'Far away, unseen, Spring faintly cries.'

This early morning on the first day is cold, clear and bright. The willows and birches of the Pools, sinuous and slim with a hint of gold, weave their image below. Beyond the hill, the sky is still a soft rose and green from the dawn's light and a wintry sun plays and sparkles on the full waters.

Today, it really looks and feels like Spring. Small birds, in overhanging boughs, awake to pay their tribute with a little song, overjoyed by the return of the watery sun. Tiny buds show on the trees; rushes and reeds that have been so bedraggled, are casting their sombre wintry covering: the first scant hazel tassels shake gently in the air.

Molehills are on the bank but no new grass grows yet. There is much activity, however. A handsome mallard, our largest duck, lifts proudly a black-curled feather on his tail. He will be thinking about a family, when he and his wife are most secretive about their nest, diving for water-weed with which to line it. Food will be carried to their young and fed from their beaks.

On the margin a pair of coots float, escorting their early family of bald babies. A sudden movement and a

shrill call from mother and, immediately, they are in a ferment, with a fluttering and flapping and skirmishing that dashes the water in sprays. Just as quickly, all is calm and the small contingent continues on its way. Any young one that strays too far from his own family, may be attacked savagely by other coots and held under water until drowned.

A moorhen feeds daintily, bobbing around for any remaining berries before flipping her wings, flirting her tail and hurrying out of sight, her feet scarcely touching the water. It seems that ducks dearly love company, for they are never far from each other.

Under the bushes, a mass of translucent jelly, with black spots, drifts. It is the spawn of the common frog, a creature that was often persecuted in the past because of its ugliness. Actually, he is an interesting creature, having been found as far north as the Arctic Circle.

I watch one, green, yellow and brown, waking early from his watery hibernation and now sitting motionless on a stone, his long, sticky tongue flicking incessantly for the insects and worms, spiders and slugs, that may be tempted by the sun to emerge. This frog, also, has to keep a wary eye open for his enemies, pike, heron and snake.

In the evenings, the Pools are loud with the croaking of mating frogs, for this creature is mostly nocturnal.

He breathes through his skin on land and water and has bulging eyes that enable him to see when swimming. His strong, hind legs help him to jump a very long way.

A little rain now pricks and stabs the Pools, where —

'Weeds a finger high,
Bow their silvery heads
With every breath of wind that falters by.'

There is little frog spawn yet. Most frogs wait until March, when a warmer sun will hatch the squirming, teeming tadpoles. After two months they will emerge, rise to the surface of the water and drop their skin and tail, having grown legs.

Across a busy road, near by, the strong urge of numerous frogs drives them each Spring and traffic is diverted to allow them to follow their instinct for food and water.

Near Merry Hill, I find a garden where scores of frogs and newts congregate at this time, round a tiny garden pond. Years ago, the natural stream here was diverted underground, but these amphibians, like rats, still seek the same source of water that their ancestors sought, nameless years ago.

Many young frogs die though they often puff themselves up in an attempt to frighten their enemies. But these are legion; large fish, hedgehogs and otters are always hungry. The frog is, however, a helpful creature to the farmer and gardener, eating innumerable insects.

Under a gilding willow, the water-weed snakes around in a brightening green and a brave stickleback, still in his wintry grey, lazes on the surface. He eats midges when they are about and larvae. The imprint of the webbed feet of ducks shows clearly in the damp mud of the verge. They play beyond, with a dipping of small heads, in sheer joy at the sun-splashed waters.

The great crested newt is not visible today or its spawn. Until late March or April, these tiny creatures hibernate under rubbish heaps or stones, for the Winter. In the mud of the Pools' floors, he waits for his mate, feeding on slugs, water snails and caddis flies. A long string of their eggs will be attached to water-weed, until the eggs fertilise.

Sadly, my little wagtail of other days is not about. Many of these charming birds have died of hunger, since they rarely accept help from bird tables and find the Winter's severity hard to endure. I miss his dipping flight and call of 'Chissick! Chissick!'

Already, in the first few days of February, I have found much stirring life and interest. I leave the Pools to their dappled shadows of blues and browns and to their movement among the diamond drops of water.

The cold of the first week deepens, but there is a healthy sparkle to the air, a slight loosening of Winter's grip. Every morning ice films the crazy paving of the garden, glittering like jewels in the sun's light.

Life is developing here, too. Ants are ceaselessly active, though mostly unseen, with an organised social order, each group having its own task to perform. Now, begins the work of laying eggs and feeding the grubs with honey.

Magically, the first yellow crocus are in flower. Daffodil buds are lengthening and magnolia buds are swelling. Green-filled leaves fan the poppy plants and the base of forget-me-knots shows a light yellowish green. Lilac buds are fresh and big, but honesty leaves are dull in colour.

There is a faint blueness on the ground where the scillas will rise and many leaves of honeysuckle now spread over the fence. More brave is the camellia. On bare boughs one daring flower opens, wide and waxen-white.

A pink haze hangs over the sycamores of the coppice, where a missel-thrush builds a nest in the bare boughs. His home of grass, moss and leaves is going to be large. I must look later for the eggs of pale blue, with red-brown spots. He seems unafraid of the magpie who watches from the Scotch pine. Strong and arrogant, this bird includes eggs in his diet of mice, snails and beetles. And yet, with his graceful flight, he is a constant delight, especially when the sun of the lawn turns his blue-green plumage to a shining jade.

Ivy flowers of pale green are appearing by the fence. Later, they will be a scource of nectar for the bees. Near the path a solitary snail wanders around slowly, searching for food. The first young, sweet, hawthorn buds are in tiny leaf, faintly greening the low, sheltered boughs of the coppice.

In the garden my daphne has not diasppointed me. Before I reach it, I smell the fresh fragrance of its tight clusters of pink buds. Daffodils are up now in long green buds that hold a tinge of gold. And what joy the full cups of crocus bring!

Near the short, thrusting tips of iris plants, a wren flies, startled from a low bush. A moment later, from a safe perch near by, he is regaling me with a loud and vigorous song.

In a dark corner of the garage, a wonderful sight — a peacock butterfly, deep in hibernation sleep. With his rich colouring, beautifully shaded 'eye' spots and his lovely shape, he is a creature to linger with. In the warmth of the sun he will fly on suede-like wings and olive-green eggs will be laid, perhaps, on my nettles. But not until full Summer will the young butterflies be on the wing.

The first, tiny, crowded buds, like snowflakes, lie on the flowering cherry. Unfortunately, bullfinches love them. Yesterday, I found this bird on my early and precious yellow-green buds of forsythia. He sat on a branch looking beautiful, with his glossy, blue-black head, warm breast and strong bill. But, he and his mate have an insatiable appetite for young green growth.

Disappointingly, the next morning brings icy patches and a bitter wind again. Few birds are in evidence, except for a quick feeding and watering by my window. A squirrel, still in full white-grey coat, scampers round. His acorns finished, he too is foraging.

Hoar frost in the second week and a still and biting air, followed by occasional snow storms. A whipping north wind drives across the garden and quickly and relentlessly, the snow falls, in thick woolly flakes that soon cover the ground. But the blue of a periwinkle is cheering, my forsythia, bravely early, waves its long strands before the wind, trailing its small gold stars until they are coated.

Over the far fields, comes the baaing of sheep and —

'New-born lambs
Within their fold beneath the hill,
Answer with plaintive cry.'

From the shrouded dampness above, myriads of weightless flakes toss in the air to lose themselves in the stillness of the earth. On walls and trees and hedges little hillocks of white appear, and an early moon in a darkening sky, rises cold and pure as the snow itself.

Morning, and well-booted and buttoned, I haste to Merry Hill. There is no merriment now but much beauty. Silent, I walk on a white carpet, pristine with frosted snow, across tree shadows faintly blue. The hills are softly veiled in snow blossoms; encrusted trees bow their branches; every hedgerow, twig and wintry blade of grass shows crystalline. In shades of white and lavender, the fields are unblemished, untrodden.

Afternoon and the snow ceases. Small birds are hidden; large birds delay their flight; only a fox, near the mating season, hungry and slinking, steals ahead of me and is gone. And I, too, hurry home to warmth and the joy of the crocus where, thrusting aside the cotton wool of the crisp snow on my lawn, they emerge triumphant and golden into the cold air.

The snows linger and are gone and, for a few days, the fields are waterlogged. But the hedgerows on the Hill are subtly changing. Blackthorn, seeming to banish the winds of Winter, is bursting leafless into white blossom, dwelling in stark beauty among its thorns. A surfeit of blossom in severe weather was thought to indicate a cold Spring and to result in bitter fruit.

The skeleton of a small animal hangs from a thorn, perhaps that of a field-mouse. It may be the victim of a shrike, who pounces and impales his prey on a thorn before eating. Patches of bright green show where the sap is rising in the hawthorn. The leaves of the beech are tiny, but pointed and gilded. Delicate as harebells, lamb's tails sway from the hazel twigs and the first rich warmth crowns the oaks. Among the speckled gold of dandelion flowers, a little thistle, blue-green, pushes through the mould of last year's leaves.

Sticky horse-chestnut buds are still small and so are the round, black buds of ash, but the elm shows a few minute, tight flowers to the flickering rays of wintry sunshine.

Deadly nightshade and nettles are starting to grow and green stars of flowers cover dog's mercury. The stinging hairs of nettles may result in a rash in humans, but the plant is invaluable to the caterpillars of the small tortoiseshell butterfly and the red admiral. The leaves contain vitamin C and iron and, when well cooked, are highly nutritious. It is believed that Bronze Age man made nettle cloth and found it durable.

With a Spring awakening, yellow gorse glows on the heath, so that one is persuaded with Daniel Defoe when he said of another heath, ' 'Tis so near heaven!' Among the thorny gorse, tender buds are opening and pale yellow shows behind the tightly closed buds of broom, which is rapidly and vividly greening. Near by, an early almond tree discloses branches of young blossom.

My little avenue of limes is a tracery of budded twigs. Here, in concealing leaves, the damp soil has induced the humble coltsfoot to be the first to open its yellow blooms and several dandelions, the 'pissenlit' of the French, glow like lamps on a wet day. More small nettles are resolutely green. At the base of shining ivy, I find my first sheltered and glossy celandine, eagerly open, and with Blake, 'see a heaven in a wild flower!'

On Merry Hill, the stark and wintry outline of many trees is beginning to soften with a blurring of tender Spring green, although the fields still look forlorn. Deep within low branches, a wee wren seems to be clearing an old nest, tugging away at a piece of dried cobweb. Buried in the hedge's ditch,

slugs are rousing to search for earthworms and the slow-worm, once known as a 'dead adder', burrows beneath the leaf mould for spiders and snails. He is really a lizard and, with many enemies, has to be cautious.

It is the third week and inviting weather, very cold but clear. Within the shelter of Whippendell Woods, Spring buds are filling on birch, oak and beech and I find a precious hidden primrose bud. A cold sun glints on new holly leaves and the wild crab is in pale bud. Young bracken stems lift, ready to curl into leaf and a sleepy bee moves among the ivy clinging around an old tree stump. In the depths of the wood, a cloud of snow white blossom hangs on the wild pear, among its delicate leaves of green.

The last week and the weather still holds. I follow my urge to go 'down to the lonely seas and the skies', of East Anglia. There is pageantry and drama in plenty here.

The water is placid under the weak, Winter sun, where erosion of the North Sea has caused the coast to be varied. Cliffs crumble, dune and shingle-ridge form and reform. In some parts, ancient forest beds have been discovered, where bones of extinct bear, deer and beaver have been found.

Here, among the dunes, the sands sculptured by the seas make curving patterns. Seedlings of the rare button-hole plant are beginning to grow and flowers will soon be forming on sea buckthorn, though few of its Autumn orange berries so beloved by the fieldfare are left. Over marshy ground, stretches of sea lavender are showing tiny new leaves and soon will come with purple flowers much sought after by bees.

I find the wetlands of the Ouse, in Winter, to be full of interest for with the flood waters come vast numbers of migratory wild fowl, among them rare and endangered species. Standing on a shingle ridge, I watch fascinated, while hundreds of waders feed. Tame mallard, teal and shoveller, the occasional white-fronted geese and numerous widgeon. Shore larks spend their Winter here, eating grass seeds and insects, while mute swans nest in the swamp.

The loud calls of many whooper swans from Iceland draw me to an even more wondrous sight, for with them are numbers of duck and waders and countless Bewick swans, who have flown two thousand miles from Siberia. Now, with up-ending bodies, they search in the waters for food.

Oyster-catchers now have white throats and live with lapwings, white herring gulls and the wading dunlin. Soon redshank and snipe will be breeding. Each bird has its own sound, for language is necessary for their survival.

'While seagulls make their ceaseless lamentations', the shining waters reflect the grace of myriads of birds, as they graze in the marshes. The air seems to quiver with a 'full tumultuous murmur of wings', as wild swans flight over flooded fields.

Evening comes.

'Silk clear the water, calm and cool,
Silent the weedy shore.'

The lagoon is floodlit and slowly the incessant movement ceases, as hundreds of contented birds settle to roost, while solitary ones, with thrusting necks and billowing wings, haste through the darkling air to their haven of sleep.

15

Changing Seas

The green seas curl and the salt waves creep
And cornflower, & harebell, the low skies sweep
In the rhythmic roll of the swelling deep:

Receding in waves on headland and bay,
The sunlight glints in silver and grey
And the breakers move in a mounting spray.

And wide and far the vast skies lean
To touch and hold the horizon's gleam
And the air is a mass of silken sheen:

But, sudden, the walls of waters loom
With a shivering light and a heavy spume –
And the world is lost in a shuddering doom:

Lost and drowned neath the water's might
As, relentless, the ocean and wild winds fight
With the untameable thrust of the tide –
through the night:

And exultant the light of the crested crown
As the vortex, inexorable, down draws & down
Till worlds in the seething waters drown.

MARCH

'Beauty walks in the woods,
And flowers from wintry sleep
Stir in the deep of her dream.'

March emerges hesitantly with the pageantry of a newly awakening world, a promise of a vast arousing of life, a surging from simulated sleep. The beauty of young life, is delicate, wide-eyed and mostly yellow; celandine, coltsfoot, dandelion and powder-yellow catkins; gorse and broom, burningly brilliant, softly golden pallor of primrose; the yellowhammer and, most precious, the yellow brimstone butterfly. In frail flight it is a promise of the coming Spring.

The severity of Winter seems to have passed and the new season appears to have 'sprung'. Yet Spring is a misleading word, for all through the year growth is developing. From Lemct, the old word for Spring, when Winter food is almost gone, comes the word Lent.

The tenderness of Spring can, however, be short-lived, for March is full of opposites. It holds balmy breezes and blizzards; anemones and ice; the crocus and bitter cold; fragile wind-flowers and frost, lengthening days and lowering skies; spontaneous bird-song and sombre silence.

The month is dominated by that ancient symbol, the egg. Even in pagan times the egg was regarded with awe and, at the coming of Christianity, used in the pancakes of Shrove Tuesday and upheld as a sign at Easter. Every country celebrates the season of the egg's significance, from the wonderfully painted designs of Romania and the Ukraine to the exquisitely enamelled eggs of Fabergé.

What promise comes with the first sunny day and 'the first fine careless rapture' of birds! What hope and inspiration! Waters stir and sing, lambs bleat, trees rustle and the sun warms all. The rites of Spring, the primeval cycles of birth, are being re-enacted under water, in the ground and in the air. It is a time when farms, forests and fields, rivers, lakes and streams, are released from Winter's trance.

But the first few days of this March are scarcely encouraging. From the warmth of my window, I see the yellow-green of honeysuckle leaves spreading below the bare boughs of the coppice, where February's faint haze of green has deepened. The lawn's grass is beginning to grow brighter and, under the sycamore, drooping heads of snowdrops hold their own quality of purity. To Wordsworth they were 'chaste', while Coleridge found them 'virgin'.

The first heavy scent comes to the cold air of the garden from untidy, resolute wallflowers, while, round my window clusters a riot of the green-bronze leaves of clematis. Where honesty is in purple flower, a few bees go about their business and tight spikes of blue-bloomed grape hyacinth make a brave border. Poppy

leaves are feathering; tulips are regal; tiny scillas in full bud as blue as speedwell and the crocus is as gold as a buttercup. And the blackcap, that 'meistersinger of gardens', has returned to his pear tree nest, near the delicate, unearthly beauty of almond blossom.

A sudden squawking from the coppice. A quarrel-some grey squirrel and a blackbird are in the pine, which is clearly part of the latter's territory. Grumbling angrily, the squirrel retreats at last in high dudgeon, to his beech.

The second day brings heavy rain and bitter cold, yet a pair of wrens, undaunted, are building in a hedge of the coppice. They have spent the Winter huddled together with others in an old nest, but this cock bird now looks most important, building a new domed nest of twigs, moss and bracken, while his wife is lining it with tiny soft feathers and making a side entrance.

18

SPRING

Meadows
fresh with dew
Robins bright & merry
Bluebells sweet and new
Chaffinch on the cherry—
Blossom after snow
Butterflies all sunning
Magnolia petals glow
And bees all a-humming
Lilac on the lawn
Flight of the swallow—
Rosy fingered dawn
And summer
to follow.

Merry Hill is saddened, with bending trees and flattened grass. Truly, today,

'The waters weep,
A stealing wind breathes in the meads, is gone.'

There is no hint of sun to light up the white band on the sparrow, but, not far away, the elegant wheatear has returned to perch on a stone.

After a while, the rain falls more softly, silent as a benediction, an offering, on the earth. It seems 'Like a private music. You do not hear, but overhear, what the rain is saying, like a child's prayer.' Every plant and bird and insect seems aware, and rejoices.

The dormouse is wakening in the bank and pigeons are wheeling in constant motion, among the new Spring crops.

I am aroused the next morning by the random roundelay of the cock wren and wait hopefully for that voice of the English Spring, the Dawn Chorus, a happening to rejoice at in wonder. But it is too early and too cold. My wren, however, does not disappoint me, with his loud, insistent song. He hops, tail in air, among the bushes, looking for food for two, now that his wife is sitting on her eggs.

As I listen, I think of Singapore where people gather at tables for a competition concert, from scores of birds in cages suspended among the trees. What a twittering, chirping, trilling and whistling must come from those sad, caged birds. Yet, our own shepherds of the past used to set a noose of horsehair, with which to catch the larks and wheatears of the heath. And, in the Middle Ages, numerous small birds were trapped, by using an owl as a decoy.

The end of the first week and the morning is brilliant, with a miraculous touch of Spring in the air, a suitable day in which to plant 'peace' roses. Bulb flowers open visibly to the sun's summons and from the coppice a pair of speckled thrushes emerge swollen with worms. Several pairs of magpies are mating there and grace the lawn in search of insects. Their voices are harsh, but they can sometimes be taught to mimic other birds. These birds seem to be steadily increasing. My magpies are using their old nest this year and may have two broods.

Even my long missing tortoise is roused by the warmth and issues forth, hungry, from her Winter home in the compost heap. She still looks somnambulant. Her life may be long, but she certainly spends a large part of it in sleep. Lingering often for insects, she ambles past the opening columbines and disappears. These slender purple and pink flowers were once called folly's flower, as their spurs were considered reminiscent of the cap and bells of a Court Jester.

The second week brings back cold but bright weather and the cherry tree is in bud. Near by, a chaffinch eyes it speculatively, while the wren's nest now holds six whitish, red-spotted eggs. Soon, the happy parents will be darting among the bushes, busily taking food to their tiny offspring.

Under the budding limes of my little avenue, yellow coltsfoot, that plant that was once called the 'first dandelion of Spring', is increasing; polished celandines, too, are opening and I find a few golden buttercups. Among the new-sprung grass of the border, grow dainty daisies and, beside them, I eagerly measure five inches with my foot. According to the old country saying, if there are enough daisies to cover this length, then Spring has finally arrived, with its small and unexpected delights.

'The petalled daisy, a honey bell,
A pebble, a branch of moss, a gem
Of dew, or fallen rain.'

Nature's joy, however, and ours is short-lived. St. Patrick's Day is bitterly cold and ends with a snowstorm. Wise squirrels, still in their white-grey coats, forage around, their secret Winter store gone.

My robin, blackbirds and thrush call briefly for food and are soon joined by a flock of glossy-feathered starlings. Noisily, they try to shuffle away the smaller birds, brawling and quarrelling, but do not attack. Clumsily, a magpie descends in the midst and, with a sweep of jewelled blue-green, the starlings wing away. Eventually, heavy winds drive them all to shelter.

A tremendous gale issues in the last week and I have to abandon my search on Merry Hill for the wild lilies of the valley, for which the district was once famous. For two days, the limbs of the trees writhe and reach, as across the Hill, the 'winds make thunder with cloud-wide wings.' In a snapping, snarling fury, bowed brambles are whipped, willows are wind-torn and hazel twigs are flung to the ground. Once, these twigs were gathered with solemnity and brought into homes, where they were believed to ward off evil spirits.

In my garden, branches float on the wind and 'in their tossing rise on a wave'. My honeysuckle is beaten, tulip petals are crushed and bushes flattened. How sad to see the snowdrops go. They seem to vanish overnight.

In long gusts the wind moans across the Hill, whistling and blowing unchecked over the bare fields. Finally, a heavy curtain of rain veils all, as night falls over the land.

Afterwards, a moist and fresh lawn breathes again; tulips lift up their heads and rejoice; rose leaves glisten; the flowering currant opens rosy buds and every leaf gratefully holds raindrops. Among the lawn's grass, several molehills have appeared, under which grey-furred animals have burrowed chambers and passages, for the mole's first new family will need a home.

Colours become vibrant with life. Swords of green iris shoots thrust upwards from the soil and I make a sudden discovery. The forsythia is ablaze with yellow stars, which light up the bare branches. Near it my blackbird trails an alien blackbird across the lawn.

Fluttering and flying horizontally, he claims his territory and the intruder flees. Whereupon, my bird opens his yellow bill and whistles, flirting his tail in triumph.

The following day, and I see a brown hedgehog. She accepts a saucerful of milk before returning to her five, tiny babies, snugly warm near the compost heap. They are deaf and blind and rely on her for food. Many babies die, unfortunately, since hedgehogs are our helpful friends.

Despite the stinging winds and slashing rains, my azaleas are in tight bud and cautious, peeping forget-me-nots have blue eyes. In the sunless air, fragrance still rises from the opening blossom of the cherry tree, but yellow poppies hang their heads, reluctant to open. I linger to listen to the song of the chaffinch, which differs slightly from that of the wild bird of the woods.

The twenty-third brings a whole day of rain and the willow of my little avenue droops and gleams golden. By late afternoon, out comes, simultaneously, a belated sun and a crowd of birds, the latter with their evening song. And a pale, fleeting rainbow shows, softly luminous. Briefly it glows and is gone. This arch of promise is still believed, by some native tribes, to be but water, glazing the glistening skin of a python.

The last few days of the month bring renewed joy in Spring, as from

'The blue regions of the air,
Where melodious winds have birth,'

my robin sings.

Hopefully, I hie myself to the Lakes, to visit my favourite hanging wood, pausing only to savour the breath-taking beauty of early damson blossom.

A solitary place, this hanging wood, a vernal wood, rising from the lake in which it is mirrored. The hills above are full of the quiet grazing of sheep where, beyond drystone-walls, patient Herdwicks gaze down from their heights. They have survived the rigorous Winter. Would they ever, I wondered, be fitted with plastic coats after shearing, or zipped boots, as in parts of Australia?

Over the peaks, white clouds scud, continually changing the crests' contours and where I stand sway golden daffodils. Softly, the wind sings and sighs in the trees, making a sweet sound like the waves of the sea. As I approach the verge of the wood, a covey of partridges takes off from the heather and two yellow wagtails fly from the reeds.

On the air, comes the rich scent of cowslips, pale green and gold and, among the vegetation, I catch a glimpse of the long, brilliant tail of a retreating pheasant, a bird which is not native to our country. The grouse that feed on the young heather shoots are hidden, but beneath an opening sprig, I find the brightly green- and pink-spotted caterpillars of the emperor moth. They, too, enjoy the plentiful heather.

Winter's mystery and shades still invade this place,

for

'Very old are the woods
And the buds that break
Out of the brier's boughs
When March winds wake.'

From fleecy clouds of wild cherry blossom, like lace of cream and green, a hawfinch with a huge bill sings his sudden song. This tree, which 'whitely welcomes Lent', is as bright as a young larch. I see an early marbled white, but look in vain for the holly blue butterflies which are increasing in the woods. However, it is too early.

Slowly, I climb through the spreading carpet of wild flowers where, among moss-covered roots, patches of primroses, wide-eyed, and violets, half-hidden, grow. Violets were greatly treasured by housewives of past years. A 'verrie daintie dish' was made into a 'pudden' which was good for 'giddiness of the heart', while dried blossoms were considered to be 'a good cure for cross husbands'.

Tiny shepherds' crooks of bracken are lifting crisp and bright and small ferns show like the horns of a young stag. The holly has glistening new leaves. Above, gold-budding beech spreads and fresh green needles adorn the spruce. A persistent tapping sound comes from an oak tree. That must be the green woodpecker, but I do not see him. Soon the jay will be building his nest, which he keeps carefully concealed. At this time, he is a great robber of small birds.

And, everywhere, under the unfolding foliage or beneath a pink-budding may tree or crab tree, are white and mauve-veined wind-flowers, blowing for a magical moment from the sweet-smelling earth. Pliny believed that they only opened when the wind blew. I watch them in delight as they sway and quiver delicately. At the first sign of rain they fold their petals daintily, a hiding place, it was once thought, for the elves and fairies of the wood.

How fragile, low and frail they are and, looking down on them from his perch on a bough, a hidden chiff-chaff thinks so too, as he sings his early Spring song.

I am welcomed home by a pair of linnets with pale red breasts and grey wings. They are preparing their nest in a dense bush of the coppice and are busy forming it with twigs and little roots and lining it with hair. I must watch. Last year they had five nestlings, from whitish eggs, which they fed with seeds and insects.

Three days left to the month and, after a drizzle of rain, we have brilliant sun and warm air. Round the garden pond, the paled petals of the cherry tree fall softly on the grass that is growing and greening. From his branch of the Scotch pine of the coppice, a keen-eyed heron waits. He comes from the Pools and has found the long Winter, when fish have been ice-bound, very trying. His hunger makes him fly long distances and occasionally he visits our garden ponds and even London's parks and lakes. This early morning, like a statue, he watches before snapping his long beak with unerring aim on a drifting fish.

Each day now brings a new and surprising joy. The dwarf quince is flowering in the small, orange-red blooms so admired by the traditional Chinese painters of long ago. The air is filled with the chatter of happy magpies and the lawn with their tail-dipping movement. At last I discover the flowers I search for, but they are not wild. Under the trees of my garden, each folded leaf half hides a sweet spray of lily of the valley.

And everywhere, in every garden, 'daffie-down-dillies' are spreading in deepest gold and crocus stud the grass. But the richness of daffodils is already becoming dimmed and tarnished, as though their radiance is too great to be sustained. And sadly, within a few days, we shall with Herrick be lamenting, 'Fair daffodils, we weep to see you haste away so soon.'

Undoubtedly, the magnolia is queen of the garden. Described by one Prime Minister as 'the chief glory of Downing Street', the magnolia was originally brought from the Far East to delight us. The sight now of the first, pink-veined buds is uplifting. Pearl-tipped, on this bright morning, they taper upwards, adding light and lustre to the Spring air. And one cup of cool cream is spreading waxen petals.

Round the Pools and fittingly for Spring, bright-eyed youth crowds, jam-jars and shrimp-nets in hand. They, too, are filled with *joie de vivre* and hopefully search for wriggling tadpoles, caddis worms, leeches and minnows. On the overhanging silver birches, they climb and hang and swing, watching in excitement the string of golden toad-spawn that floats on the water.

Under swaying lamb's-tails, new grass of an exquisite green is growing and more dandelions are opening. Foraging in the soil, I find a tiger beetle, his tiger-like jaws always ready to snap at his prey.

A great flapping of wings and two herons lift over the trees where they roost. Is there a sedge of herons near? I watch as, long of leg and yellow of beak, they make their ponderous flight with trailing legs and upturned toes.

On the still waters minnows move with the breeze. The kingfisher's young love them. They are fed to them by their parents in their nest, dug deep in the soft earth of the bank and lined with fish-scales and bones. The colourful kingfisher, nomadic through the Winter is now settling down with his family. A grey wagtail, with a bobbing tail, suddenly darts over the water after flies. Near by, a pied wagtail runs in small jerks, his head and tail moving rapidly. He is even more skilful in the air, manoeuvring quickly, for his clamorous nestlings. A quick pause to snatch at invisible insects and he disappears.

Pussy willows, as soft as kitten's fur, and gold-dusted sallow, once known as 'goslings' and used on Palm Sunday, grow. Silvery catkins sway and brown catkins hang and, among the new-feathered plumes of the greening undergrowth, a lean water vole nibbles. His hidden nest of woven grass must be near.

Among the patches of drifting, emerald weed, where the waters are quiet, cold and dark, much life is beginning to stir. Water beetles and horse leeches are full of activity and perch and pike grow fatter. The latter, prolific, can lay thousands of eggs. A tiny spider, red underneath, shows black against the pale green of a young reed. Carp, introduced from China, are stirring from their hibernation in the Pool's deep mud.

Scores of bull frogs are preparing to leave the water for the land. Food is still scarce so they often eat one another or are swallowed whole by grass snakes. Strangely, they are safe inside the snake until they reach its stomach. Moved by the same strong instinct as toads and herds of zebra, these frogs follow the very path their ancestors have taken for thousands of years, in their search for water and food.

March is waning and the pleasant weather persists, with cool winds and cold blue skies. Was it this month that made Beethoven declare, 'When all else fails there is always the countryside'?

On the heath, the heather is blooming and early bees are in attendance. With every movement of the air, scattered harebells delicately droop and sway. The common sowthistle, eaten as a vegetable in the Middle Ages, is almost in yellow flower and gold is showing on the ragwort. Dog's mercury grows among stones. This is supposed to be a sign of ancient habitation. Shepherd's purse is tiny and shaped like the upturned leather pouches once worn by peasants.

In the shade of one of the stones, two male adders with rough scales and a strong sense of smell, have emerged into the sunshine and are climbing around the stone. Having shed their skins, they sometimes dance together to display their strength. The stronger will mate with the female, who will retain her eggs until they mature.

Bracken is growing taller and crisply curling in tender fronds and the broom bush is richly yellow with flower, the slightest warmth of the sun bringing out its heavy scent. Here, I find my first brave butterfly of the year in flight. It is the yellow brimstone, visiting flowers to sip their honey. Being often the first of the year and as yellow as butter, it gives the name to all butterflies.

In the fifteenth century, butterflies were thought to be birds or 'smalle fowle'. This yellow brimstone has hibernated among thick ivy and slept through the snow and frost of Winter. I watch with delight its careless free flight and its fragile beauty. It was once called Yellow Bird and has daintily shaped wings.

What a useful bush the gorse is. Many spiders and ants make it their home and the yellowhammer loves it. Here, the female bird builds a cup-shaped nest and lines it with horsehair. I heard his song, but did not see him, as he repeated, 'A little bit of bread and no cheese'.

A sleepy hedgehog, cream and brown, slowly emerges from the shelter of the gorse. A captivating creature, with his pointed face, small eyes and long legs, the hedgehog has been around for fifteen million years. A Preservation Society has been formed in order to discover more about him. He finds his food by his sense of smell and the movements he hears below the soil. He looks awkward, but yet can climb up ivy. This one is interesting to follow, as, hungry, he ambles about for slugs, beetles and birds' eggs, being not averse to adders.

Across a wide space I stand to watch 'the low-eared hares upspring from cover', among the grass and herbage from which they are feeding. Suddenly motionless, they seem to freeze, listening with ears extended and sniffing the air. Then, in rapid, incredible speed, they circle round and round, the doe darting away and the hare flying after her and leaping over her from side to side, as they run.

Once the hare was called Puss and was revered by the Celts, who used it in a ceremony to foretell the future. The movement of my two hares is the very essence of Spring, as they soar over the earth with agility and grace, fleeing like shadows through the whining wind.

And above the trees of the Pools, against the blown clouds, come the clamorous wings of

'Swan's maiden flight, in the climb
To a tremulous harebell's crest.'

Spring Winds

The wild and whipping winds are gone
And, suddenly, and unaware
The magnolia tree is ringed with bloom
Oblation-lifting to the wondering air.

But yesterday, for one sweet hour,
In regal beauty glowed each flower,
And now the pearly petals lie,
As floating lilies, neath the sky:

One last long look
              let adulation rise
And mingle here, magnolia wise.

APRIL

'The year's at the Spring
And day's at the morn,
Morning's at seven,
The hillside's dew-pearled.'

The pageantry of Spring comes to its first blossoming in April. It is now that the half-hidden life of Winter is revealed; in sweet, fresh scents; in soft-hued colours; in joy expressed *sans cesse*. Prodigal, profuse, the miracle of Spring is manifest in the abandoned movement of youth and its teeming life. It is a time when, after the seeming inertia of Winter, one is confronted by the mystery of life and made to marvel at the magic of its renewal. A time when children, in a fever of Spring delight, make cowslip bells, daisy chains and dandelion rings or plait rushes and chant rhymes.

'For the Winter is over and gone,
And the time of the singing of birds has come.'

What are the heralds of Spring? For some, it is the cuckoo that tells us 'Summer is icumen in'. Yet the willow wren and the chiff-chaff often return before he does. For others it is a powdery, swinging catkin, a first fragrant violet or shining celandine. Or is the first frail butterfly the true harbinger of the season?

Often when Spring seems firmly established, however, variable weather returns with winds of March and cold rain of February.

This April of 1981 is no exception. The first few days bring bitter winds, cold showers and sharp frosts at night. The magic however, is unmistakable, for the next morning the Dawn Chorus greets me, heralding a day of intermittent sun. In a joyous outburst the feathered songsters of the garden vie with one another. Robin, chaffinch, wren, thrush, trill and warble and,

above all, comes the sustained song of the blackbird. Surely it is his voice that weaves the pattern of sound into a tapestry of exquisite melody, the occasional discordant note only serving to accentuate the harmony of the whole.

Another unfailing joy; the old magnolia blooms with undiminished beauty, lifting its half-opened cups of delicate pearl regally to the grey skies. Veins of pink show on tight tapering buds and, in the slanting light of the reluctant sun, the ivory petals glow with a faint flush of rose.

When I venture forth to discover more secret gifts of April, what a welcome from my friendly chaffinch as he greets me with his song 'To meet you! To meet you!' Sometimes he calls 'Pink! Pink!' Offering crumbs, I go carefully to inspect the nest of grey lichen and moss which he and his mate are building in the apple tree. Soon they will both be kept busy feeding and rearing their young.

Everywhere the tender greens of Spring show and the rockery, despite frosts, is a blaze of colour, with yellow of alyssum nestling among the royal hues of lobelia. Blue-green iris leaves pierce the soil like scimitars and gaily coloured anemones with blue scillas wave their petals in the breeze.

Primroses, in shy posies close to the ground, peep from crinkly green leaves, truly the primrosa of the Spring. Their bright leaves are believed to relieve rheumatism. The other day I was shown an unusual primrose with several blooms on one stalk and near it, in the same patch, cowslip bells grew. Polyanthus is the result of this abnormal growth. The 'primmie rose, eldern berrie and dandielyon' wines were much loved by past generations.

From the coppice comes the deep note of a wood pigeon and there is much raucous activity where the rooks are building in the high tree-tops near the farm. Though they are clannish they do not hesitate to steal nesting material from a neighbouring rook's nest. I hope that they will not invade the coppice at the end of my garden, where several pairs of magpies have settled. These are handsome birds and often walk, proud as peacocks, over the lawn. Small birds, while keeping a respectful distance, do not seem unduly worried by them. Slow and heavy in flight, they are content to stay near their large, domed nest in a tall sycamore, which they have made for their few, blue and speckled eggs.

Blue tits seem to have deserted us. Perhaps they will return in the Autumn. Last year they made their home in a hole in the wall for nine, white, red-spotted eggs and fed the young ceaselessly with numerous greenflies.

Under the old limes of my little avenue, where the buds are slowly breaking, celandines gleam and glow and dandelions are abundant and beautiful. Their leaves are edible and much beloved by Canada geese. The flowers, when fried in butter, are believed to be even more nutritious.

After the Winter ploughing, trees everywhere are of vital importance to birds. Winter's ice and frost has broken, releasing many insects and small animals. Food is easier to find and every male bird marks out his territory by tree, gate or hedge.

My song-thrush, who has been missing for some time, is back. Usually he is rather timid and keeps in the background, but could always be wooed in Winter with currants and sultanas. Now he has chosen a wife and she is happily making a nest in a fork of a spruce. She is carrying grass, roots and moss and soon I shall expect to see four or five brown, blue-speckled eggs.

Along the path I follow the slimy, silvery trail of an awakened snail. 'You glide in silken silence,' says the poet and rarely is one seen. The trail is quite long but there is no sign of the snail. When the larger snails are abroad, my song-thrush watches for their trail and tracks them down, banging the hard shell against one of the numerous flints in the soil. He then enjoys the juicy morsel inside.

Though snails can live for months without food, it is surprising to read of a sea-snail which has, recently, survived foodless for three years. A large, yellow and white-lipped one, it normally forages on seaweed. After being stuck on to a decorative sea shell, this particular snail, destined for posterity, has survived, being eventually released into its natural habitat.

The winds of this first week have gone, bringing a suddenly beautiful day, one of April's surprises; a day precious in its softly coloured pageantry; a day to be savoured and not to be missed; a day on which I see my first vivid butterfly on the wing. It is a small tortoiseshell, out of hibernation and fluttering round the leaf of a stinging nettle, under which she will lay her eggs. When the caterpillars hatch they will spin a silken web over the leaves on which they feed. This vivid, tawny butterfly likes the gardens and homes of people.

After lingering with the magnolia, I turn to admire the blue of forget-me-nots, the rosy buds of flowering currant and the early glory of the cherry trees. Blossom seems suddenly to be everywhere, lighting the world and deepening the blue of the skies.

But the buds are wise not to open, for the next day brings back the cold winds, followed April-wise, by a brief spell of sunshine.

Merry Hill, this morning, is indeed merry! It is a symphony in green and white, filled with swelling buds, opening flowers and joyful outpouring of bird-song. Tiny red stamens spring from the tips of hazel buds, male catkins shake golden pollen, bronzed beechbuds taper and brambles thrust eagerly outward. The horse-chestnut is cautious; only a few buds show, buds of dark varnish, sticky to the touch. The tree looks somewhat damaged after the prolonged frosts.

My sycamore, also, seems reluctant to burst into leaf, though buds are pink and plump. But the wych-elm flowers and sallow and willow will be ready for Easter crosses, for which yew boughs were once used.

A field of buttercups, like a sea of gold, is a sight of wonder. From segmented leaves, that can blister the hands, the little open cups gleam, polished and deeply yellow. As I walk, their pollen powder stains my feet.

An occasional bee is about by the wayside, where white stars of stitchwort grow without visible support. Deadnettle flowers and that unusual plant, lords-and-ladies or jack-in-the-pulpit, grow and I find my first crane's-bill. Wild strawberries, the *fraises du bois* beloved by the French, have a few white flowers. Tall Queen Anne's lace, not fully out, is like fallen snow. These plants attract small flies which carry pollen to other plants.

An early orange tip, that delicate and fragile butterfly, visits garlic mustard or Jack-by-the-hedge, whose heart-shaped leaves have a smell of mustard when rubbed.

There is much 'cuckoo spit' on flower stems in the lane. This is a larva. Inside the froth, which is made up of air bubbles, are the blue-green froghopper nymphs. Until they emerge as adult bugs, they need protection from predators.

The hedgerow, happily left uncut on Merry Hill, though erected for man by man, is an ideal retreat for the abundant life of Spring flora and fauna. Hawthorn, in particular, offers a good refuge for birds and their young, its twisting stems making it doubly secure for tiny birds such as wrens. Vivid green and frilly leaves are now appearing and among them a willow warbler, one of the earliest of returning birds, sits and sings his sweet Spring song. I stand apart to listen.

The guelder rose is showing posies of white, yellow-centred florets and, almost hidden, I find a few pretty pink blooms of ragged robin. Near them, an upturned stone discloses earwigs, black and crawling and active. They are busy now making nests, often in those of old field-mice, where they store pollen and honey for their grubs.

Across my meadow, which is forever a source of delight, the cuckoo calls. Although they are lazy and heedless birds, refusing to build a nest or rear their young, their cry is the epitome of Spring. No doubt, this cuckoo is looking for his favourite food, woolly caterpillars. At one time his advent was the signal for Cuckoo Fairs.

The damp air of this second week is today full of sun-drenched scents of grass and weed. Dog's mercury is beginning to bloom. It was believed that the Greek god Mercury was the first to discover the healing properties of this plant. Near flowering bird's-foot trefoil I catch a glimpse of a lovely common blue butterfly, while among a patch of annual meadow grass, a wall brown waits.

By the path, patches of bird's-eye speedwell are bright, like clouds of blue fallen on the grass. Pink heads of yarrow and delicate lavender of lady-smock are opening. The legend says that St. Helena found 'My Lady's smock' growing in the manger beside the Virgin Mary. Glancing, blue-eyed speedwells were thought to promote a safe and quick journey. They certainly seem, at times, to follow one's path.

The old barn, with its orange tiles on its sloping roof, looks empty of hay, as I pass. Its inhabitants, the bats, are not visible today. Wrapped in their wings, they lose much body warmth in the Winter as they hang from the rafters. This month one baby will be born which is usually blind for a while and clings very tightly to its mother. Rats and voles and even fish are caught and also moths in mid-air.

Among the tiny plants, centipedes, worms and caterpillars are rousing. Violets have to be searched for. Cautious and shy, they nestle half-buried in dead, dried leaves. Charles I enjoyed a drink made from their heart-shaped leaves, to which sugar and lemon had been added.

In looking for them, I almost fall over a brown-curled ball. It is a hedgehog only half awake from his Winter home in the ditch. Having eaten he will doubtless stay hidden until these April winds subside.

Not far from cattle in a field, I find two or three rare cowslips, deeply yellow and fragrant. Once they were called the 'keys of heaven'. Almost hidden by grasses, white clover grows among its trefoil leaves, which were thought long ago to indicate the Trinity. How children love to search for a lucky, magic, four-leaved clover, which is fancifully supposed to grow only in Ireland.

With the celandines have come the returning swallows. A charming legend says that cranes, on their long flight from wintering in Africa, carry the swallows for part of the journey on their strong wings. Here, on Merry Hill, they turn and roll above me, in a graceful diving after insects. In their beaks they have taken mud and clay from the banks of the stream and mixed it with dry grass for their nests under the barn's eaves. Feathers from the farmyard and last year's coils of lamb's wool are skilfully woven. Pale, red-speckled eggs, one each day, are laid. Now, on a ceaseless quest for food, they skim low, with darkly blue plumage and long, streaming tails. In May, the swifts will follow them.

Hovering over the meadow flowers I find a yellow brimstone butterfly out of hibernation. Its bright wings with four orange spots are fragile, but its flight is free, a sure sign of Spring.

Apart from the early manuscripts of monks, Dürer was the first realistic illustrator of these small and ancient plants.

On my way to the Ponds, and my little avenue, arched with limes, is a joy to see, each new leaf unfolding in a brilliant green. It is the middle of April and cold winds have returned but the Pools, with their

margin of flowering sallow, poplar, and tasselled silver birch, are a hive of industry. Pussy willows are big and catkins droop and trail over the water.

Frogs, mice, voles and insects are wakening. Water ducks weave patterns on the surface or dabble, diving with webbed feet and upturned tail to the bottom for food. A few wide-open flowers of bright yellow show on the silver-weed at the water's edge, where a moorhen prepares to lay her eggs among the softly shading rushes. A sudden flight and she almost runs over the water, disappearing with a throaty cry.

There is a constant movement on the surface, where water beetles and snails are busy. Whirligig beetles create their own small ripples, tiny silver fish move in small waves, but the caddis fly, waterproof, stays on the surface. Pond skaters, with only their feet touching the water, move quickly. This is the month when the male attracts the female by agitating the water and stays with her until their eggs are safely laid. When a suitable spot on a submerged rock has been found, about thirty are cemented on. Within three weeks the young come to the surface, where they run the risk of being swallowed by an adult.

On a warm stone a young frog rests from seeking insects and suns himself in the brief rays of light. He should be more wary, for green-gold pike 'hang in an amber cavern of weeds', waiting, and, hovering in the shallows, a perch is hidden.

A water shrew spends his time between the bank and the water, in his endless search among the duck-weed. The Winter grey of stickleback is turning to a bright blue. The weeds lie in long ribbons and, weaving between them, each fish swims in its own solitude.

Thin scatters of intensely blue forget-me-nots grow and tall bulrushes with a hint of orange in their brown. Standing motionless at the edge, a heron, with soft grey plumage and yellow eyes, waits and watches for an unsuspecting vole to come within reach of his darting, sharp beak. At this time of the year, when mating, this elegant bird turns scarlet.

A handful of swallows swerve and cleave the air for their last meal on this April evening. As the winter-like sun sinks red and stormy, the evening primrose unfolds and the heron, well fed, lifts away above the trees in a long-trailing flight to the nearby reservoir. Flapping slowly across the flushed skies, he leaves the smooth unruffled waters of the mere to its own dreams and its own reflections of shimmering silver trunks.

And soon, almost unnoticed in the heavens, is

'The little silver moon that April brings,
More lovely shade than light,
That, setting, silvers lonely hills
Upon the verge of night.'

In three weeks of cold winds the month has given us only a handful of warm days. The avenue limes, that yesterday held out new leaves of polished emerald, now hang dead-lustred and listless. In this cold, dull weather small birds seem to be hidden, saving their songs for brighter days. Although the coppice is greening and thickening there is little evidence of the magpies, but the damp lawn is studded with daisies and my evening blackbird never forsakes me, whistling his tune companionably near my window.

Even at Kew the cold wind tosses the blossom like the manes of wild horses, softening the colours of the dark-winged cedars. Down by the lake the water is whipped into tiny, restless waves, over which weeping willows, feathered and golden, lean low. Under their shade and shelter the swans have hidden their nest of twigs and dry grass. Mother swan sits on her five eggs, rearranging them fussily from time to time. While mating and rearing, the parents will keep close to each other and their babies.

A dab chick sails by, taking two of her brood for their first view of the cold but exciting world outside and a Canada goose, with her six plain goslings, wanders on the verge, savouring the greenest of grass.

In the fork of a partly submerged branch is a most intriguing nest formed of bits of blue rag, silver and red paper and intertwining twigs. What enterprising parents!

A common mallard comes close to wait hopefully and greedily for crumbs and a little grey wagtail bobs about, showing the yellow of his breast. Dipper birds, blue and white, fuss round for food. Here, many wild ducks will come at night to drink and feed.

Lacking their warm plumage I retreat to the lily house where, in cosy tranquillity, dragon-flies hover above water-lilies of lemon, lavender and rose. With long, trailing stems they float above their green-blue pads. On a cushion of green, one lone lily of pearly pink sits in isolated loveliness, contemplated by several humans and one appreciative frog.

The sudden shower has ceased and, outside, the pale sun grudgingly emerges. Instantly, the trees resound to a choir of birds. It is Spring and the blossom has bloomed. Despite the weather Kew has its bower of cherry trees. Heavy laden, the boughs of snowy blossoms fall around me like the slow waves of a scented sea. With one solitary duckling I linger to savour the fragrance.

But this capricious April is not over. In the darkness of night snow falls and only humans are taken by surprise. Some country roads are closed with floods and many telegraph wires are down. Trees, grass, fields and flowers have a new covering. The world is white and crystalline, with snow below the boughs as an added blossom.

Happily, it is April snow and soon disappears. After two days of rain this changeable month gives a full offering, pouring all its beauty into the last days. My garden is radiant and with Thomas Browne I gaze 'thro' a window faire and comely' at 'delightful borders and heare the ravishing music of pretie Small Birds.'

Exultant buds open wide; butterflies, perhaps the most lovely and graceful of all living things, are on the wing; the plum, beloved by the ancient Chinese, is in bloom and lilac plumes scent the air. I find my first spray of hawthorn, that garland of cream fragrance considered by Victorians to be unlucky, despite the fact that earlier centuries held it to be a symbol of fertility.

Readily, the yellow brimstone butterfly returns to the broom and the air is filled with the urgent, wheeling flight of house martins. Necklaces of jewelled flowers on the clematis frame my window, a mass of pink and shining beauty in the sun. The sycamore has at last cast its rosy sheaths and from the lawn comes the delicious scent of new-cropped grass.

And in the woods,

'The beech is seen
Undeniably a queen,'

shading a first patch of flowering bluebells, a foretaste of next month's wonder. The world seems filled with joyous music and a faint mist lies on the hills, lending an added loveliness. And, suddenly, a rainbow arches in the sky, transluscent, softly-hued, as April in a shimmering veil says farewell.

Spring Remembered

A mauve rain falls from the blossom trees –
& slow-falling snow drifts down on the breeze;
Heavy and laden their burden they give,
With a grace & a beauty that forever will live;

O, lift high your crown now
    triumphant with bloom,
Then, soft and translucent
    return to earth's womb,
For the Spring, weary-waited,
    has given her all;
While petals of almond,
        magnolia, fall
With cherry's warm hues
& the broom's palest glow,
O, linger a little,
    and love, ere you go;

This rich Spring, so prodigal, fair & profuse,
Will brighten and burgeon in memory's use,
In moments of bleakness, of blindness, regret,
Of all Springs remembered, the loveliest yet!

MAY

'The chaffinch sings on the orchard bough.'

May is still a magical month, if not in its beginning a merry one. A panorama of gentle pageantry, this early May morning was heralded by cold and squally showers. Now, new and fresh, the sky is pearled and a frail mist hangs over the valley.

Merry Hill! What an appropriate name for a lane along which to wander in May! It is indeed 'a little and a lone green lane', like that of Emily Brontë, where 'the darling buds of May', christened so many times, have now triumphed. Deep in the hedgerow the magic of May crowds in on every side. The month has burst forth with everything that the season can offer. Under its bower of arching trees, damp and fragrant, the whole lane has a quiet charm.

'A refuge green and cool
And tranquil as a dream.'

After the rain everything glows and birds sing with renewed joy. Trees throw a maze of dappled shade. The horse-chestnut is lifting tight rosy buds. Trailing flowers from fans of leaves cover the sycamore. Young beech leaves are radiant and the oak is breaking in yellow-green foliage. Wrens and robins rejoice and bees murmur in the limes.

From the banks, flower faces peep upwards; cow parsley heads are white with bloom; shy daisies cling to the verge; the white nettle flowers show and dandelions glow, deeply golden. Like forget-me-nots, speedwell has a glancing beauty and stitchwort is studded with stars. May blossom hangs heavy festoons of cream and pink. The lane is a symphony in green and white; the air heavy with drenched blooms.

'O, see how thick the kingcup flowers
Are lying in field and lane,
With dandelions to tell the hours
That never are told again.'

A warm, hazy day and on my way to the meadow I pause to admire a wild, Spring garden, with its mingling patches of harebells, violets, lady-smocks and primroses. Whether the flowers are 'thrum-eyed' or 'pin-eyed', primrose paths, as in the Isle of Wight and Devon, are wondrous ways. W. H. Davies understood, when he exclaimed, 'O, rare primrose, born when the Spring winds blow!'

A ripening spray of sorrel holds a brilliant small copper butterfly, his wings iridescent, fully awake to the joys of Spring. His caterpillars will feed on the sorrel. What magic butterflies hold with their free, fragile and evanescent beauty! Down the centuries man has regarded them with awe and affection, some tribes still holding them as a symbol of fertility.

By the five-barred gate I linger to see the richly green valley below, where Herkomer and Turner so often sought and found inspiration for their paintings. For Merry Hill must surely have been the scene of many festivals and numerous rejoicings. Probably the village maypole, earlier regarded as a phallic symbol in the age-old fertility rites, was raised here and the green hill resounded to cries of merrymaking and music of country song and dance. A perfect scene for the simple pageantry of May Day processions, when carts and drays and horses were dressed in garlands of woven flowers and plaited grasses. A time when small daughters, aspiring to be chosen as May Queen, echoed the words:

'Call me early, call me early, Mother dear,
For tomorrow will be the happiest day of all the
                                    glad New Year:
The maddest, merriest day!'

The thin shriek of a tiny shrew distrubes my thoughts. I move and, running round in circles, he disappears below brambles and broad coarse leaves of dock. Outflung rose briers have new leaves of pale green yet seem reluctant to break into flower but the bittersweet is in flower. A few blooms of herb Robert hold a richly deep rose and the gnarled boughs of the apple tree has delicate petals of pink and white. An early honey bee, not yet fully awake, is visiting them busily. In the time of St. Augustine the apple was considered to be sacred.

Merry Hill is merry indeed; ringing with bird-song, gay with flowers; sweet with scents and alive with pulsating movement.

Down comes the rain and my meadow is abandoned. Two dull days and through the mist my hopeful clematis, which had in April lifted myriads of petal faces to the sun, now hangs dejected, each yellow eye hidden. But a sudden gleam glances and poppies of orange have burst forth in a humble, cheerful flamboyancy.

The end of the first week and the air is softened, yet still fresh. Lesser periwinkle is studded with mauve-blue flowers and azaleas, in all hues, are ablaze with blossom. And reigning over the garden with a pristine loveliness, blooms of the old magnolia are resplendent. Calling incessantly, sadly, collared doves seem to have invaded the land. Thunder rumbles in the air.

One beautiful day in the second week, and the clouds darken. As though seeking refuge, small birds shoot across the sky. Drenching rain follows, leaving a glistening world, each separate rose-leaf glinting with drops. Light glows on petals. Silver pools lie in the sunken circle where the stone girl dabbles her toes. Across a sky of harebell blue clouds of cotton wool move slowly.

The high-pitched song of a blue tit comes, 'Blue-blue! Tit-tit-tit!' He seems to love my sweet-smelling honeysuckle, which has been partly broken by a marauding black cat. The pretty songster has something to sing about. In a hole in the wall I discover his nest. A work of wonder; a weaving of lichen, wool and the webs of spiders. Moss cushions nine tiny eggs, white and red-speckled. Mother and father will be kept busy feeding their young with numerous caterpillars and greenflies.

Meadows hold magic. I stand to absorb the special atmosphere. For surely meadows are unique and fast disappearing. The air is warmly scented by growing, moving things, for meadows have a fragrant, fragile beauty all their own. Beyond, the young corn is rising, pointed and green. The clear song of a wren cuts through the air and a skylark hovers above his territory.

Tall feathers of waving grass are breaking into tiny flower, perfect and scented. Buttercups, yellow rattle and the honey-fragrant charlock grow. A few scabious, mauve-blue, make a foil for the yellow and white. Sweet-smelling pink clover show among horse-tail and rye grass, on which small children still lovingly count:

'Tinker, tailor, soldier, sailor,
Richman, poor man, beggar man, thief.'

Foxtail and quaking grass have the delicate tremble of harebells. Over thistle and mauve-white flowers of blackberry, a wall brown butterfly flits. She has laid her greenish eggs, one by one, on blades of annual meadow grass, where her caterpillars can browse and grow.

Common mallow, originally used in May celebrations, is tall near the gate, but the pink trumpets of lesser bindweed riot gracefully, clinging to stronger plants and demurely closing its fragrant flowers at dusk. The meadowsweet, though not yet fully out, is already claiming its title of 'Queen of the meadows'.

'The flowers of the field have a sweet smell:
Meadowsweet, tansy, thyme,
And faint heart pimpernel;
But sweeter even than these
The silver of the may.'

In vain, I search carefully for one wild cowslip, that bloom so nostalgic and fragrant. How rare they are, although one railway bank in Hertfordshire is still thickly covered in Spring, each flower head nodding gracefully and trembling its gold. Once cowslips were so profuse that each child could make a sweet-smelling posy and every mother could take pride in her home-made cowslip wine. Few butterflies are around. The weather is too cold for them. Also for the greenish grass snake which lives here and likes the sun in which to bask. This snake was revered by the Ancient world. It is harmless, continually flicking its forked tongue for insects.

Under the ample eaves of the old barn, martins are busy building nests of mud. Swifts are ceaselessly on the wing, catching minute insects as they wheel. It is amazing, the thousands of miles these birds wing in migration, outdistancing any other bird.

I pass the Japanese cherry tree, which is laden with pink bloom and to which two chaffinches come each year on the same day, foraging among the copper leaves and buds. After her lean Winter feeding, the mother bird is building her strength for her new family. These colourful birds appear to love ash buds best of all. Their nest is probably in the coppice, although one year they built in my thick forsythia. In the wild they can live for seven years, but in captivity, safe from predators, they may survive for twice as long. The anxious parent bird hovers near the delicate nest of lichen and cobwebs which holds her precious fledglings and, keeping one step ahead, hops before me.

Between the trees of the lane the sun now filters in palest gold, falling on darkly purple bugle, on meadow saxifrage and patches of brilliant new green moss. Among the snowy layers of blossom that reach out from the hawthorn boughs, a drowsy bee flounders.

Meadow Magic

Blue green the young corn is rising
And buttercups gild the green verge,
But where are the meadows, the meadows,
Where our youthful memories surge?

We wander through grass, long & trembling
Where kingcups and lady-smocks gleam,
We capture the whole of earth's glory
In a reverie – or is it a dream?

In the dew of life's morning we lingered
Where angel-eyes scattered the grass
And meadow sweet, milk-white, caressed us
And never the wonder will pass!

And sweet were our thoughts & our fancies
And bright were the long summer days,
As the skylark soared from the clover
While we wandered the meadow's cool ways.

Tread gently, the violets are hidden,
Pale primroses offer you gold,
And campion pink and blue borage
Our faltering steps enfold —

And green are the fingers of chestnuts
That hold aloft candles of rose —
And will all the magic of meadows
Be lost, where the water flows?

Smaller than the honey-bee, the wild bee in its gold-striped, black velvet coat, is a sure sign of Spring.

The quiet lane is a tunnel of greenest gloom, bringing

'All that's made
To a green thought and a green shade.'

I linger in this shadowed world of green and white. In itself, green is a magical colour, as Mary Webb knew so well when she exclaimed, 'I find a hundred shades of it in one field!' Green seems to be the epitome of new life, new growth. One typical English village, in an ancient ceremony, still celebrates the end of May by presenting their newly gathered green boughs to their nearby cathedral. One of the many country customs that combines pagan superstition with Christian belief.

Returning, I see the oak now shows tiny curled leaves; ivy creeps forward with fresh green tendrils and new tips of holly are vivid and glossy.

May is now in full swing. It is the middle of the month and the morning is greeted by that true harbinger of Spring, the Dawn Chorus. How wonderful is this first concerted song of the birds. Slowly, the darkness diminishes and the stars, like pale lamps, fade. Tremulous and hesitant, the chorus begins, swelling and blossoming to a crescendo of sound, a paean of praise. And, following, come sudden flurries of activity which are not to be ignored.

What an offering, suitable but totally unexpected, for May! I discover a tiny cowslip bud in my own

garden. How I shall treasure it!

My friend, the blackbird, ceases his frantic song and hops near to welcome me, his head on one side. Clematis, in all its morning splendour, has thrust upwards and outwards in a wealth of pink profusion to cover the south wall. Azaleas are now fully out in shades of rose and red. A large white butterfly visits several flowers, lifts, and is gone. Sadly, the blue of forget-me-nots grows faded and tired, but poppies are still rampant, popping up everywhere in splashes of yellow and orange.

Desultory rain for a few days. Far away, thunder rumbles and retreats. Sky turns a dull indigo and all bird-song is stilled. Through the unnatural gloom, I see the yellow-green tassels of the sycamore and the rose of their dropping sheaths. The willow weeps in earnest, its long fronds already a duller green. My large orange

tulips are depleted, their petals like frayed flags, torn by the wind. But the rockery still glows in blues, violets and yellows from aubretia, lobelia and alyssum. And, though limp, the lime leaves are vividly green.

After the leaden skies, a glistening world. From the coppice comes the 'Chip-chop-chap' of the chaffinch, showing a glimpse, as he flies, of soft yellow and brown. And over the fields floats the call of an unseen cuckoo. Too lazy to build, is she seeking the nest of a hedge-sparrow into which she will leave her solitary large egg?

The middle of the month and it is decidedly un-Maylike, bringing wild, wet and windy weather. My sycamore tosses its clusters and fig-like leaves. Being of the maple family it gives good shelter to birds. Beneath it, tawny coloured toadstools spring up overnight on the saturated lawn. From the mountain ash in the coppice comes the clear, repetitive cry of a gentle song-thrush, most welcome in this strange silence.

More rain follows, bringing a deeper glow to leaves as they fill with life and growth.

A break in the weather and I haste expectantly to Whippendell Woods and

'The dim dells where, in azure, bluebells dwell.'

The canal, with the blackness of its far flung cedar, is quiet by the old water-mill and its miniature waterfall. This, with its scintillating and falling cascades is a pageant in itself. A huge blackbird with a glossy, orange bill watches, enjoying the perpetual spray. Here flowers of watercress bloom in the running water and, hidden near by, an adder lurks, the zigzag on his back hidden by the undergrowth. Unless attacked he is not aggressive and has his own enemy, the badger. Recently it has been found that the adder's venom can be useful in curing circulatory diseases.

Summer

Golden sun and sea & sand
A ladybird within my hand:

River's reach
And rosy mallow,
Warblers darting
Neath the willow.

Wonder wings of dragonflies
Flash of white when lap-
wings rise:

Lavender
Valerian
Convolvulous
Delphinium:

Butterflies upon the bower
A marigold – a gillyflower,

A frail and fragile
Swallowtail:
Then moonlight &
The nightingale.

The woods are a delight, filled with bird-song and a shimmering vision of bluebells. Beneath the trees the air is like wine, with a slight breeze and intermittent sun. Blossom smothers the crab apple and a petalled snow showers from the wild cherry. Boughs of beech hold out layers of tender green, with trailing, copper-coloured bracts. In the dim woods, the horse-chestnut has belatedly burst into fingers of brilliant green, where rosy flowers are opening. Soft gold shows on the oaks but the ash seems reluctant this year. On tapering twigs, birch saplings have delicate foliage and conifers rise tall and dark and straight. Through the boughs splashes of golden sunlight make radiant patches of Spring grass.

And, beneath the trees, cool bluebells, scented. Between wine trunks and grey-smooth boles, swathes of them cover the ground in drifts of blue and mauve, pink and purple. Like a fallen sky that has gained richness in falling, they glow. Round an ancient tree stump they lie in shaded curves. And, close by, rose campion and herb Robert daringly lift their heads. And all through the woods the bluebells grow, reflecting the sun in pockets of azure light, their bells seeming to peal for joy of the Spring.

It seems strange to think that these flowers were once used to stiffen the ruffs of the first Elizabethans. Now, with their drooping bells and rosettes of leaves clinging to the ground, they are a symbol of May, gladdening the eye and the heart with their beauty.

Through the silver, polished trunks of the beech comes the derisive cry of a green woodpecker. His long, thin tongue probes incessantly into bark, seeking maggots and grubs for his young, which are safely concealed in their bottle-shaped nest.

Swallows swoop through the glades for insects and a dragon-fly flashes by. Once, the latter was known as the 'Devil's darning needle' or the 'flying dragon'. For a second, he seems more dazzling than the sun, with the light flashing on his green body and gauzy, slender wings, as he darts in rapid motion for flies and tiny moths.

My veronica is out with spikes of white and mauve flowers, but the garden's chief glory now is the clematis. A riot of twisting stems and eager blossoms, it spreads lovingly, lifting elongated cups of pink to a watery sky. In a splash of brief sunshine fat, purple-red peonies are cautiously unfolding. And, suddenly, a large white and a yellow brimstone are on the air. How sensitive these butterflies are to the sun's warmth.

Despite the dull drizzle of the next two days, London pride, or lad's love, shows fronds of dusty pink and the wisteria still offers its scent. My blackbird, too, persistently chases away any interloper from his territory.

And, unaware, Spring is fully upon us. The innumerable buds of the coppice have swelled, changing its Winter darkness to a reddish warmth. Sprays of birch and hazel show swinging yellow catkins and the old elm tree holds a nest of chattering starlings.

Willows are green and silver, having lost their gold. With a strident squawking, several pairs of magpies are building big, untidy nests. Undaunted by their noise, my blackbird comes close to me with a sideways nod of the head, as though to say, "Good-morning!"

Delighted to find my first ladybird resting on primula petals. Am tempted to hold it in my palm as when a child, and solemnly recite:

'Ladybird, ladybird, fly away home!
Your house is on fire and your children are gone!'

Always, the insect departed hurriedly, no doubt in search of more aphids and greenflies. This old rhyme apparently originates from the annual, Autumn burning by farmers of tree debris and leaves. In the Summer the ladybird lives on the tree and its harmful insects. Being unable to fly, its young were often trapped in the flames. The Winter is spent under stones or in cracks of houses. Known in medieval times as 'Our Lady's bird', its Latin name being connected with its seven spots.

May is drawing to a close and is still cool. Too cool for my lilies of the valley. In their sheltered corner they are quite sparse this year. No squirrels have appeared since April. No doubt they are busy rearing their young in the dreys of the coppice. From a branch of lilac a small wren, with upthrust tail and vibrating body is pouring out a wealth of song. These birds are forever with us. Flying low and direct, they run around under the thick azaleas like tiny mice. Despite their loud song they are rather captivating. Their wee nest of moss and fern, lined with feathers, has seven glossy white eggs speckled with red-brown.

A welcome sight. My first rosebuds have appeared. Their cream-yellow petals are almost bursting. The last waxen magnolia flowers still linger, with a thrush for company, under their pink and wine-stained cups. Pale thrift is starlike, but wallflowers are faded and bedraggled. Yellow rain of laburnum is blooming, yet peony buds are still tightly sheathed.

More and more, the garden flowers and flourishes. And everywhere, plumes of lilac, white, purple and rose, scent the air. A colourful tortoiseshell butterfly hovers over silvering discs of honesty. Cascading stems of clematis have bronzed leaves. The first spears of iris with long buds cleave through the ground and Solomon's seal shows broad, smooth leaves. Remnants of tulips flaunt their frail petals and, on tall stems, columbine hangs dainty heads of cream and purple.

The last Sunday in May brings sun, watery but sustained and a brief song from the thrush. Isolated claps of thunder and small birds hasten to flee for shelter. Boughs shake restlessly. The bright sky of early morning is broken by black clouds which hang over half the sky. Heavy rain, and the heavens are emptied of bird-song.

The storm has passed. Beyond Merry Hill a veil of mist rises from the fields and a yellow light clouds the lowering sun. The last fragile petals of poppies droop sadly, until a fresh breeze dries their fluttering. I discover a tiny cowslip in bud in my border. What an offering, suitable but totally unexpected. How I shall treasure it!

Deceived by the capricious weather, a young thrush, a robin, little perky wrens and the ever-present starlings, their wings sparkling like jewels, come to feed.

A dull day, but greater bindweed riots in abandon in the hedges, with tiny black insects crawling inside each trumpet flower, looking like black threads among the pearled white. Hedge parsley is in full bloom, lifting cream parasols among the thrusting brambles. Honeysuckle is still fragrant.

On a gooseberry bush I find a magpie moth in black and white with orange on its underwing. A small skipper butterfly flutters past. He feeds at night and will soon be hibernating.

Among the profuse growth of the hedge side, wild oats which birds love, are swelling. Chamomile grows, with its golden centre and turned-down petals. Because of its faint apple scent, this is named earth apple by the French. From it, oil is extracted for medicinal purposes.

Poppies are scarce on the Hill. Their seeds can remain dormant for many years, as they did on Flanders fields, before blooming in scarlet patches.

A day of torrential rain, causing much flooding. The following afternoon is quiet and Merry Hill subdued, under a pearl-grey sky. The may, weighted with water, hangs low. Lime leaves sparkle and holly glistens. A pretty common blue butterfly, venturing, perches daintily on the pea-like flowers of bird's-foot trefoil. On the verge, a splendidly bespeckled thrush pecks hopefully in the shaggy grass for rising worms.

And, from his nest deep in the meadow, a lark with a warm breast and a tufted head, rises to the serene of heaven. Up and up he soars and into the sun, this blithe spirit, offering his melody of joy to the retreating month. Truly, he sings of the exaltation of larks.

JUNE

'All the live murmur of a Summer's day.'

June holds all the pageantry of early Summer, the tranquil quietude of river and valley. 'Slowth', an East Anglian word, is expressive of its atmosphere. The early, excited business of birds has softened. Spring flowers have gone and Summer blooms have not yet realised their fullness. In shorter nights and longer more radiant days, the former dampness is no longer apparent.

This is a season of scents, which have become profuse and varied, 'the delicious fragrance' that W. H. Hudson knew so well.

In spite of a lessening in the Dawn Chorus, June is also a month of joyous sound.

'There's music in the sighing of a reed,
There's music in the rippling of the breeze.'

This early June, however, brings dull, warm weather with prolonged thunder at night and the light wind of the next few days develops into a gale. For a week strong winds persist and skies grow dark and songless. But, like the movement of the sea, these winds become more measured.

Poppies glow in the garden and clematis still riots around in abandon. Occasional gleams of sunlight on tight spikes of delphinium which are slow to unfold and on peeping blues and yellows of lupins. More iris bloom, flag-like, but honeysuckle flowers wait for warmth.

A cold persistent wind and the roses are reluctant. Sweet pea plants slowly beginning their clinging climb, but purple flowers of honesty have gone, leaving pearly discs of fruit.

During the desultory rain of a grey day an odd burst of song from a lone wren and out comes the full sun as though in answer, along with a cluster of young starlings to forage on the lawn.

Pleasant warmer weather and after the recent rains, the garden gleams, the sun lending every leaf a silver light. Buds of rambler roses are now eagerly swelling.

A brilliant dawn to the fourteenth, bringing out *tout le monde*; butterflies, birds, insects, roses and people. Valerian is in rosy spikes and delphiniums, having lingered so long, now burst into the glorious blues of stained glass. Flowering dogwood is in pink blossom with four floral bracts and red-green leaves. Mock orange and lavender are fragrant and a violet mist covers the ceanothus.

Like the poor, house sparrows are always with us, building their untidy nest in the drainpipe. Straws protrude and only half conceal the five, pale-green eggs.

Bushes of azaleas glow in all shades of warmth, but forget-me-nots are seeding. Their tender blue eyes will be missed. A few solitary stars of clematis still cling to my window.

A great whirring of wings and a collared dove, swollen and gorged, its white tail fanned, lands heavily on the lawn. Around him, for once in unison, gather blackbirds, robins and little wrens. Refusing to be intimidated, they harry the dove until he withdraws to the coppice and the full creamy flowers of elder. These birds seem to have increased in the conifers. Finally, my blackbird alone is left, looking highly satisfied, having jostled without attacking, for sole control of his domain.

Two dark, dark days, silent of bird-song. On the low stone wall my robin sits as usual and waits expectantly for food. But worms, busy underground, do not appear.

June may have flickered faintly but it has not yet flamed. Through a midday dullness that is almost like dusk, the blue and white bells of campanula gleam and orange marigolds light their lamps.

In their familiar site by the garage, a second family of blackbirds is established. Father blackbird has been raiding young lettuces in order to get at the insects underneath.

Privet is in scented flower and virginia creeper is moving at a prodigious rate, its tiny tendrils of green and wine-red groping forward over the brickwork.

A sudden heat follows on the twenty-second. Although belated it is much appreciated. Early morning, and 'a stealing wind breaks in the meads'. Merry Hill is full of 'live murmurs', where the sound of running water is pleasant to hear. The trees are darkening with leaf and tiny acorns form on the oak. Birds and insects are busy. Belated limes are flowering and many bees, as if to make up for lost time, are humming frantically.

Midday and the far fields shimmer in a haze of heat. The stiff and upright ears of corn have the green-gold sheen of early ripeness.

My meadow, still unshorn, is a sight of loveliness. Sheets of buttercups in the sun's light are like molten gold. Billowing in the slight breeze, grasses are tremulous and scented. On the verge a few creamy feathers of meadowsweet sway. Here too, are faintly pink spikes of greater plantain, slender tendrils of vetch in purple flower, ragged cow parsley so irresistible to insects and spear thistle beloved by goldfinch. With their bright plumage, friendliness and song these birds are rightly known as 'a charm of goldfinch'.

One small meadow is already low with hay, scenting the air with its fragrance. In another, knapweed, yellow rattle, ox-eye daisies, spotted orchis and quaking grasses are all falling before the relentless machine. From a birch tree comes a 'Chiff-chaff-chiff'. Flying from the Mediterranean the chiff-chaff has been here since March, being one of the earliest birds to return. Of soft-hued browns and yellows, he builds his domed nest of dry leaves, stems and moss, in a dense bush.

On purple spear thistle, a painted lady, *La Belle Dame* of the French, has settled. These graceful, tawny-orange butterflies are most adaptable. Soaring from North Africa they have been found within the Arctic Circle and in deserts. In their migratory flight they move at a speed of fifteen miles an hour. What a wonderful instinct and perseverence these fragile creatures have, sharing their sense of urgency with pigeons, salmon and turtle.

In the hedgerow, goose-grass has a thickly hanging mass of pale green flowers and lengthening ivy tendrils partly hide red campion. Purple crane's-bill and rose of musk mallow grow near. A lovely marbled white butterfly moves in slow flight over cat's tail grass, into which the female has dropped her eggs and on which her caterpillars will eat after dark.

How quickly the may blossom fades and disappears. Flowers of six petals show on the prickly brambles, but the exquisite filigree of Queen Anne's lace is less white, less clear. This plant is named after the mother of the Virgin Mary.

As I pass the old barn, which is already almost full of hay, a friendly robin goes in jerky flights before me. This barn is the home of a small colony of bats which are, in England, a dying species. These winged acrobats fly at dusk in search of baby snakes and small mammals. They have an unusually wide wing-span and make a squeaking noise in flight, having one baby a year and hanging upside-down to sleep.

The day has changed to one of full Summer and petals of late blossom lie like drifts of confetti. On a sunlit stone a wall brown butterfly, quick to seize advantage, takes her zigzag flight and stays still, absorbing the warmth.

The hedgerow is a miniature world, sweetly tangled with the heart-shaped leaves and trailing blooms of bindweed; of byrony and wild guelder rose. How delicate the flowers of bindweed!

'What frail, thin-spun flowers
She casts into the air
To breathe the sunshine and
To leave her fragrance there.'

Red stems hold pink buds of wild columbine; cream nettle flowers flourish and star-heads of the greater stitchwort gleam white. This plant was once thought to cure 'the paine in the side'. Blooms of the herb Robert are named after the Abbé Robert, founder of the Cistercians. Sprays of wild rose overhang smaller plants, offering on the same branch flowers of white, pale pink and rose.

June is in its third week and Suffolk is smiling as I pass through level fields heavy with grass and scattered with flowers, those 'luxuriant meadow flats, sprinkled with flocks and herds' that Constable loved. It is an ancient landscape, where a stone found near the river has been proved to be two hundred million years old, being swept down from the north in the Ice Age. With its towering churches and great windmills, this country is also a landscape of charm.

The day is perfect. Into drowsy airs above me, candles of horse-chestnut lift and scented lime flowers hang. The noisy waters of early Spring have calmed and the river winds its shallow, slow way under low-flung arches of elder, wind-leaning willows and ragged lines of grey-green poplars. Dwarfed and gnarled by the gales of Winter, some are still ruggedly root-clinging.

In shady places the yellow of balsam with long, spurred flowers grows tall and here:

'The stream
Lies calmed to a brown dream.'

Flecks of golden light dance on the water and flash on the wings of minute insects. It is quiet. No trout are seen rising in the wondrous curved leaps they keep for the early morning and evenings. Badgers, too, are rarely seen, since they prefer to drink at dusk. In February four or five cubs are born, having their own underground chamber in the set. The mother keeps this very clean bringing new leaves daily for their bed. They are very active and playful and are fed with grubs of wasps and bees. The adults feed mainly on snakes. Recently in the south, badgers have been seen near houses, but they are usually shy and country-loving and very cautious when leaving their sets.

Under the willows by the water's edge, a redstart builds. Hoping for an unsuspecting, rising fish a heron, that patient and clever fisherman, glides gracefully downstream. A hungry owl watches perched and hidden, for mice that scamper below.

A sudden movement of a stick-like body and a dragon-fly, having waited for his wings to harden, flashes by. After two years in the mud and slime, he has become a bright, eager insect with glancing, slender wings. What a brilliant transformation and so short-lived. Incredible to think that such a delicate creature can move at a speed of thirty miles an hour.

The longest day of the year and light until late in the evening. The middle reaches of the river are green with iris leaves where a few yellow flags wave. Mayflies are on the wing enjoying the sunlight. Thick rushes grow with fine pointed stems and hairy stalks of comfrey hold drooping flowers of pink, blue and mauve.

On a reed stem below the surface a damselfly has laid her eggs. The bright blue of a kingfisher shows through the foliage, a shining fish caught in his strong beak.

A little breeze stirs the aspens, turning their leaves to silvery discs, but the high reeds are moveless. Common loosestrife is tall and marsh marigolds with goblets of gold, cast a burnished glow on the waters. These have been flowering since March. From the shadows comes the 'peep-peep' of a dab chick.

Scores of tiny flies speckle the surface and a little flock of minnows drifts blissfully with the slow-moving current. The hard-working and sharp-eyed heron darts and there is more food for his gaping young, waiting in the shallows under the shade of the marigolds. The delicate winged caddis fly is unseen for he flies by night. A sudden splash and a water-rat dives from the bank for his fishy meal.

Midsummer Day and the pleasant weather is broken by north-east winds, but only briefly.

In the upper reaches of the Stour the otter has been sighted. This captivating animal lives in his holt and has learned to swim and hunt under water without gills. Though it has a piercing cry, it can slide silently into the river, trailing an arrow-ripple behind him as he moves.

On a cool, bright day I come to the Otter Trust. Twelve cubs, born here in the ideal conditions of captivity, are soon to be released. By tracking them through the dense reeds, where they eat moorhen, baby rabbits, fish and eels, it is hoped that their declining numbers will be increased. With his intelligence, his fascinating whiskers and his almost human use of finger-like feet, this nocturnal animal is growing extremely rare. Perhaps next year we will catch a glimpse of this delightful creature swimming by the river's shady bank.

Pass a brilliant field of wild mustard, looking like an upland cornfield. Beyond it, a sweet scent rises from the bean flowers.

What a strange beauty have the stony wastes of Breckland, where Neolithic Man mined for flints. Through the woodland rides and ranks of tall fox-gloves, sight a roe deer, a small species, short antlered whose reddish coat turns grey in Winter. In this month his one or two kids have been born.

For many years the stone curlew, known as the Norfolk plover, has come to the open wild places here to breed. Foolishly, these birds nest on the ground but are well camouflaged, being big-eyed, long-legged and mottled with browns.

In contrast, the great grey shrike of the heath,

known as the 'butcher bird', is a rare and solitary predator. Like a small hawk, he has black and white wings and a magpie-like tail. Perching on bushes and watching for small birds and rodents, he impales his prey relentlessly on a thorn before eating.

At the Fritillary Reserve is a wonderful sight. A whole meadow has the rare blossoms of snake's-head with their accompanying butterflies. These purple or white lilies are no longer found in the meadows of the south-east. With their drooping cups they are one of the rarest of English species. They grow in damp places that are mowed annually.

Found a wild and magical garden with an indiscriminate mingling of scents and colours. Here nettles are encouraged and weeds left unhindered; the beauty of butterflies and other insects is of paramount importance. Herbs are cultivated for usefulness and seeds from desolated meadows are gathered and distributed to all parts of the world.

Beyond the long shingle ridge of the beach at Orford, the marshes are extensive. Below the Castle the River Alde passes Havergate Island, a most interesting place where the graceful avocets are bred. After a hundred years these waders have returned again to nest here and at Minsmere. They are rare birds with slim and long, upturned bills.

Flying over the marshes in June and July, the uncommon and wonderful swallowtail butterfly comes, its green and black caterpillars feeding on the milk parsley that grows here. It is as beautiful in colour and grace as the swallow after which it is named and rarely found anywhere else in England.

Behind Aldeburgh, tall native reeds grow in purple and brown plumes. Among them the bearded tit finds a home and insects for food. These colourful birds are known locally as reed pheasants.

From far away comes the boom of the unseen bittern, that patroller of the marshes. This mysterious and rare bird has nested here for centuries in the reeds and sedges of East Anglia. Being almost extinct in the nineteenth century, it has now happily been re-established. Tendrils of a mauve vetch, the marsh pea, cling and climb up the reed stems. This plant is almost confined to these regions, and attracts small caterpillars which feed on it.

In the grassy places, little red harvest mice are busy in continual search for seeds and insects. Here, they skilfully build a round nest to balance on the strong reeds, where they must be always wary of owls and weasels. The latter, which are good swimmers, are also on the look-out for water-voles and moorhens. Another enemy of the mice is the thin and supple stoat. The playful baby stoats are known as kittens. Being smaller than a rabbit, the stoat can worm its way into a burrow. But, having scented his prey, he usually pursues it relentlessly, until the exhausted rabbit appears to become mesmerised and is killed by one ferocious bite. The stoat's Summer coat is a reddish brown which in Winter turns completely white.

On the edge of the marsh, a stonechat, which is not often seen, has been busy building under the sheltering thorns of a gorse bush. This bird with a dark head and brilliant breast has a short, dark tail and mostly stays in this country throughout the year. His name derives from his song, which has been described as the sound of falling pebbles.

Above the stonechat's nest, flies a golden plover, whose precious eggs may not, by law, be taken from the nest.

Among the dunes of Aldeburgh, are the wiry leaves and spikes of white flowers of marram grass. The prickly sea-holly will have bright blue flowers in August. These attract hover-flies, bees and butterflies; also the six-spotted burnet moth which flies by day.

The little tern comes here in April, making his vulnerable nest on the shingle, where sturdy sea peas grow. The purple-red blooms and blue-green leaves lie low among the stones in brilliant patches. In the famine of 1555, the local people were thankful to eat the fruits, regarding their profuse growth of that year as a miracle.

Much later, in October, the shore larks will arrive to spend their Winter here, feeding on grass seeds and insects. They have a distinctive tuft on their yellow and black heads, but are unfortunately decreasing in numbers.

Over the green swards and soft-hued hollyhocks of Southwold comes the call of the cuckoo. Near by on the shore found a grotesque and strange sight for this pleasant month of June. Dead oaks thrust drowning arms upward. These, I am told, were destroyed by the salt floods of the encroaching seas. Further inland, hedgerow oaks have outstretched limbs and trunks completely clothed with thick ivy. These are known locally as 'ivy todds'. When the ivy flowers, many bees, butterflies and moths are attracted. The 'todds' are also a refuge for numerous birds of the thrush variety, who feast on the berries of Autumn. The male purple emperor butterfly loves to hover around the tops of these oak trees.

In the pools of the dunes lives the rare natterjack. Known as the 'running toad', he has a yellow line down his back and his eyes are golden. Mostly hidden, this amphibian scuttles about at dusk in search of food.

Beyond the marsh, shimmering in the haze, a sudden sun lights up the silver of sea purslane and the marram grass of the dunes. Lapwings, weeping and wailing, turn in the air above large brown heads of swaying reeds. Bright yellow patches of stonecrop grow, with horned poppies, burnet rose and pink and white of marsh mallow. From the white roots of the latter plant, marsh mallow toffee was once made.

Near the coast, bunches of seven-spotted ladybirds made a strange sight, clustering closely on reeds, where a sedge warbler sings sweetly. Known as 'the poor man's nightingale', these birds, 'Clinging so light to willow twigs,' build their low nest around supporting stems. They winter in tropical Africa. After a storm and much feeding, they cross the Mediterranean and the Sahara in one arduous flight.

On the glancing surface of the pools, a wild goose anxiously launches her flotilla of young, while from the margin comes the croaking of the first frogs. Low over the marsh Kentish plovers fly, with black bills and trailing legs. A long-legged heron stalks slowly before lifting, in heavy flight, with a 'Krark! Krark!'

Here the reed mace grows tall with black-brown heads above its blue-green leaves. Hidden in them a pied wagtail has fashioned his nest. In the pine tops a beautiful hoopoe waits for crickets, scorpions and grubs. His large, black-tipped, fan-shaped crest is decorative, but his harsh song is far from attractive.

Deep in the marsh the swamp spider, the largest in Britain, preys. Walking on the surface of the water, he warily spins behind him a line to the safety of the land. The peat pools at the source of the Waveney are its sole known haunts.

These marshlands, whose existence is continuously being threatened, were drained in part by the Romans, a wider reclamation being made in the seventeenth century by a Dutchman. As I watched, a mysterious wind crept and stirred among the reeds, breaking the eerie silence and a lone bird swept in high circles above the solitary wastes.

The last week of June and distinctly cold, with a persistent wind. Constant chirruping from the birds. Are they trying to keep warm or are they protesting? There is little sign of their babies, whose nest must be a warm refuge.

Roses, to my continual delight, are not so shy. Many are in full glory in the garden and lupins are warmly corn-coloured. Tall, heavily laden foxgloves, unabashed, shake their bells in the breeze. A wee wren darts in quick movements under the azalea bushes.

'A tiny, inch-long, eager, ardent,
Feathered mouse.'

The last day brings light winds, with belated sun at noon. On Merry Hill it silvers the leaves of the trees, where horse-chestnut holds high, full and sweetly smelling flowers, its long-fingered leaves spread for light and space. I found a tiny scar like a hoof-mark below each bud.

In the hedgerows there is a sudden surge of new growth. Parasols of flower-clustered cow parsley form a lace-like pattern and the verge shows the yellow and red of 'eggs and bacon' or bird's-foot trefoil.

From the sweet freshness of Merry Hill I turn homewards. Beyond the trees:

'The sun's last beam leans low'

and the garden now holds the quietness of evening. On flame and white, purple and violet petals, the lowering sun casts a glowing, golden light. Like pale ghosts, moths flit around my head and, across the valley, sounds the long hoot of an owl. Through my world of green comes the sleepy twittering of birds, in 'a charm of singing', that Francis Kilvert knew so well. And from the sycamore, through the softened air, lifts my blackbird's evening song of praise.

VALE

Walk where the Stour wends its winding way,
Meadow-meandering,
        through the lush green verge;
Where willows lean down low to touch
The water's deep, pellucid calm;
And may boughs trail their milky weight
Through grasses sweet and cream and pink
With clover and with campion:
By yellowed lime and sober elm,
Where lifting flowers of chestnut spread;
And, drowsy with earth's loveliness,
Gaze through the bronze of burnished beech
Alone, while the vale's last light
Glows palely on the—
        crown of Dedham's tower,
Lifting and rising, as a diadem.

JULY

'That which the universe
Lacks room to enclose,
Lives in the folded petals
Of this dark rose.'

The rich profusion of June has, in July, brought airs hot with thunder and heavy with the season's scents. Meadows hold a languorous fragrance. Breezes play in the mellowing stalks of the cornfields. Golds have deepened. Woods have darkened their green. Bird-song is less eager; bees and insects busier.

But, if June is the month of roses, July brings them to their full splendour. For surely roses are the essence of an English Summer. Long ago, the word 'rose' was used for any flower. Today, roses hold the beauty of all flowers within them and speak eloquently for themselves.

Soft yellow blooms encircle my window: deep red ones border my lawn. Bushes are bronzed or thornless: floribundas abound. Pure white roses are many-budded, are subtle of scent. Birds are attracted by their glossy green leaves. And the evening air adds to their loveliness, deepens their fragrance. For

'No one knows
Through what wild centuries
Roves back the rose.'

Warm and sunny. Merry Hill is flooded with sunshine. A visit from that colourful bird, the jay. He probably has his sharp eye on the oak in the coppice, but the acorns are not yet fully formed. Squirrel-wise, he sometimes buries these under long grass.

The meadow still holds its beauty; a warm and scented shade is creeping over it, mostly from sorrel, which is not eaten by cattle. Bordering trees still in fresh leaf. As with flowers, the names of grasses have a fascinating ring; timothy grass, hedge-bedstraw, sheep's fescue and tufted hair. The quaking grass quivers. All now are gay with flower or heavy with seed. In the lane, tiny moths 'tremble in a shimmering grace' and float on.

Earlier in the year, caterpillars of the cinnabar moth have bred in the meadow, especially liking the poisonous ragwort. Bees are at work among the white clover of the verge. How cows love this flower — contented cows 'in clover'.

Verdure has flourished in the hedgerows. Lilac of field madder shows. The uneven petals of heart's-ease have a red mark among the yellow. Vetches are purple and gold. On the grass of the margin, yellow rattle grows, with coltsfoot that was once used by the Romans as a cure for asthma.

Against the warm earth of an adjoining field, a small patch of mustard burns brashly. Once, these

51

pastures around were called 'heartlands'. Strange to think that our wheat of today comes ultimately from the grasses that grew many centuries ago: from green foxtail, graceful cheat-grass and clinging goose-grass. Since the fifteenth century the latter has been fed to geese.

Over the meadow, an orange-tip butterfly flits in and out of the sun and shade, circling in the same path before it finally rests on the head of a lilac lady-smock.

For eight days the horizon is heavy with heat and all life basks in the sun. Hedges are wreathed with the unearthly beauty of convolvulus or bindweed, that plant beloved of day insects and night moths.

The scent of wild roses lingers on the air and elder is in bloom. Delicate pink petals are held on the thorny, out-stretched and arching stems of blackberry. Wild honeysuckle, yellow flushed with red, winds and twines about the bushes. It is waiting for the darkness when it will be visited by night insects.

My meadow revisited. It is sadly shorn of its little homely plants and looks bereft as though it has lost all its beauty and life. A quiet stillness, a haziness hangs over it. No butterflies flit, no ladybirds charm, no sight of insect or grasshopper. Where thick, luscious grasses and dainty florets delighted the senses, now lambs lazily graze. Life has not entirely forsaken my meadow.

Much of its growth seems to have been transferred to the verges of the lane, for there is a sudden surge of viridescent growth in the hedgerows, which seem full of movement.

A loosened stone and centipedes stream out to pour in ceaseless motion over the damp earth. Mullein throws out yellow spikes that are taller than foxgloves and St. John's wort and milkwort are flowering. Close to the soil insects hum and buzz and, deep in the bank, a wasp visits his nest, where shepherd's purse, tansy and speedwell bloom. Deadly nightshade hangs purple bells and a few last violets hide below pale yellow foxgloves. And among them little white butterflies move aimlessly.

Where trees crowd into a shade, cows stand motionless, only their tails restless against flies. Above them, beech leaves hang, a rich viridian, and lime falls listlessly.

But if movement is drowsy below, the air quivers with song. From the coppice comes the monotonous call of the collared dove, the coo of the wood pigeon and the harsh chattering of magpies, while a grey squirrel swings noisily and effortlessly from branch to branch.

Between the crevices of an old wall grow tiny herb Robert and the star flowers of stonecrop, that food of the rare and beautiful butterfly, Apollo. Near by, grow the rounded rosy heads of wall spur valerian.

The fine chestnut on Merry Hill that was damaged by the April frosts is belatedly showing new fans of leaves. In the ditch is red campion, that flower that always offers a fresh, moist smell. Almost invisible, a grass snake coils through the undergrowth. There is no sign of her babies but they cannot be far away, hidden in the shade and wound about each other for comfort.

The second week in July and heralded by beautiful weather. Garden brimming with movement and sound, where one

'Can hear all day long
The thrush repeat his song.'

Visits from my wary robin and a strange, white feathered blackbird, with a scattering of sparrows. Interesting to read of albino blackbirds, finches and martins. It is believed that their white feathers are the result of artificial feeding, rather than the insects and worms which are their natural food.

A red admiral butterfly shows brightly coloured against a seeding teasel head. He is also attracted to buddleia, the 'butterfly plant' which is visited by many insects. Almost hidden, the lesser periwinkle hides among vivid leaves. The red admiral, whose caterpillars feed on the nettle flowers of the lane, was once known as the 'red admirable'. His swift and graceful flight is a delight to watch.

For five minutes the little pond is left unnetted and now a goldfish floats sadly on the surface. Is a magpie responsible or that great catcher of fish, the long-necked heron, that was sighted in the dawn?

The dry lawn is being invaded by creeping, purple self-heal. Dainty boughs of silver birch droop lifelessly, but gnats dance ceaselessly and a pair of faithful bull-finches feast happily on seeded poppy heads. A solitary tortoiseshell visits the fading polyanthus, where purple and white heather grows brittle under the sun.

Delicate trumpets of fuchsia hang in the heat and petunias are most elegant, flaunting their pink and puce petticoats. The leaves of flowering dogwood reflect the rosy warmth of the flowers. Marigolds blaze. In the eighteenth century, 'mariegoldes' were used with 'sweet bassel' in a stew and were also useful for a curative poultice.

The early evening cools a little, the last of the sun casting long shadows. Beneath the bunched and golden keys of the sycamore, my blackbird pursues his mate over the grass, where honesty is turning to shades of jade and wine. As I watch, a toad with large red eyes and flashing tongue emerges from his Winter home in the hollow of a tree stump. He has stayed there sleeping, but must seek the pond of his birth in which to mate. Across the lawn he prowls his ponderous way, seeking caterpillars, slugs and worms.

The ninth day and a yellow light persists on the horizon. Hazy skies of gold and grey follow, with rumbles of thunder. An uneasy air moves in the coppice, stirring the trees in a ceaseless murmur. Their shadows move restlessly.

More rumbles of thunder and the air is heavy and hot. For three days there is little sun. Tiny pink-tipped daisies, fresh faced, appear among the green of the lawn, but their cousins the marguerites look coarse and harsh. Buds of lemon, rose and mauve wait on the hollyhocks.

The second week and blue-black clouds presage the elusive thunder. After a chilly day, much sun and wind, but no rain falls. A capricious breeze carries away dandelion clocks and on the air a small flight of fork-tailed swallows looking purposeful, as though already thinking of migration.

Between cracks in crazy paving, pin-cushion heads of mauve scabious appear and caps of candytuft look decorative against the grey stone. Petunias thrive in the heat, but goose-grass has twined overnight around a precious rose bush. Amid the humming of bees, a stickiness falls from the linden trees of my little avenue. My friend's bees have swarmed. Her grandmother always solemnly insisted on informing her bees of any births, marriages or deaths in the family. These insects are certainly clever in locating the approximate position of the sun. Interesting to read that royal jelly, the substance that revitalises queen bees, has been found to be beneficial to humans.

Only a few violet, wide-petalled flowers on my late clematis this year, but the Madonna lilies are about to unfold and warm-coloured sweet-williams exude their scent. In their border, too, tiny begonias are waxen and cheerful.

Two wrens are bobbing about and feeding from the ground. Cautiously, they have made several nests but the one in which they are rearing their young is special; it is lined with feathers. For no apparent reason, a sudden tapping on my window. It is one of the magpies. Is he attracted by the sun's reflection on my enamel bowl which is brightly gold?

The twentieth of July, breezy and warmer, followed by swift and welcome rain. The parched earth rejoices. On Merry Hill a golden gorse is like a burning bush, illuminated with myriads of sparkling drops. Small flowers release their fragrance. Bees seek nectar in the warm markings of the petals of 'love entangled'. Like clover this plant, known also as bird's-foot trefoil, is enriching the soil of Spring. Minute grass heads, fine and delicate, tremble joyfully with every slight murmur of the fresh, damp air.

Two dull days with more rain and my garden is still. 'I hear leaves drinking rain', a thirsty, unheard sound. Roses sparkle and glimmer again.

Thunder and sudden lightning in flashing flames of brilliant yellow, violet and rose.

The garden breathes after the storm, but it is chilly for July. My blackbird turns from his inspection of the lawn and its succulent treasures to welcome me. The twilight 'is dim with rose'. Across a tree bole lies a soft, five-fingered leaf shadow. The sweet scent of many flowers hangs on the air. Pale shapes of roses rise before me as I walk, in smudges of gold and white. The garden is slumbrous, its colours muted, in the last unruffled calm before nightfall.

Standing among the roses and breathing their fragrance, I remember other gardens loved. The roses of childhood with soft-scented petals and shining leaves, sweet to touch above the dark earth: a sixteenth century garden, where wisteria twisted and writhed in the agony of age, sending forth heavy cascades of lavender flowers gilded in the sun's light. And, beneath it, a stone cherub admired forever a magnificent magnolia.

Always remembered, is a garden formed and designed by an artist for dreaming. Each curving way brought a new delight and every arch of stone or bough framed a fresh landscape. And, on the lawns, a peacock stalked, spreading his wonderfully wrought tail, with its eyes of Argus. His wife, the peahen, hid her drabness in the undergrowth, near a nest she had made for their children.

In the evening, when the peacock's strident cry was stilled and darkness brought a dreaming, scented silence to the countryside, came the sad song of the nightingale.

'She sings, I hold me peace.' Over a moonlit pool, the 'delicious notes' lingered, sweetly insistent, clear and glittering as a sudden white light.

'Our dreams are tales
Told in Eden
By Eve's nightingales.'

The last Sunday in July and oppressive. The woods are dark and sombre. The air seems to be holding its breath, waiting apprehensively for thunder. A party of partridges, fat and almost tail-less, disappears hurriedly through the undergrowth. Among the debris of a tree stump, a fierce looking stag beetle gropes.

The small sound of bluebottle. He is afraid of wasps, who have been known to bite them in two pieces, flying off with one half to his larvae and returning instantly, with unerring instinct, for the other half.

Under the conifers, common enchanter's nightshade, pale pink, grows small and upright and rose bay willow herb is rampant. Bluebells, having lost their blue, hold green seed-pods.

A bullfinch has forsaken the orchard and returned to rediscover the woods. The shining satin of beech leaves has softened and darkening foxgloves have an air of mystery in their inner bells. Hairy bodied bumble-bees are visiting them and carrying pollen. A useful medicine, digitalis, is made from their leaves. These woodland foxgloves rival rhododendrons in their beauty.

A thing to treasure lies on the ground, the blue and black-barred feather of a jay. From a hollow log, where a blue tit has built a nest, comes his high-pitched call, a clear song of 'Blue-blue-blue! Tit-tit-tit!' Like the other tits he has only one brood of ten young, which are now learning to fend for themselves.

An angry, growling noise from above, where a grey squirrel is active in the tree-tops. He swings into view, bright-eyed, with twitching tail. A small and brave chaffinch challenges him, furiously ready to protect her young, and the grumbling squirrel retreats, going after easier prey.

Silence returns and from a bush a wren sings.

But this July of 1981 is not yet over. Crowning its summit comes the pageant of the WEDDING.

The last week dawns clear and radiant. Even so, anxious eyes watch the skies, for this is to be a unique occasion, a time when the nation's affection and ingenuity is added to Nature's bounty.

All is well. The twenty-ninth brings a warm and golden day for a warmly golden celebration. It is a time of perfect unity in ceremony and tradition.

In the darkness before dawn, the skies proclaim their pageantry; beacons have lit up the encircling hills, flashing the message to the whole world. Behind the façade of the Old Buckingham Palace, a firework picture of the Prince of Wales and Lady Diana shines in a great whirling ring of fame, amid the triumphant Vivat Regina of the State Trumpeters.

Now, on this glorious morning, comes a Pageantry of Rejoicing which the most elevated and elegant in the land share with London cockneys, Scots crofters, Geordies and Welsh miners. There is a magic abroad, released by the complete and utter friendliness of sharing.

Beacons burn, bells ring, guns fire, Morris Dancers dance, fancy dress parties revel and people frolic in fountains. And all under the blue and scintillating sunshine of an English July, with trees in full foliage, bright and shining in the Mall and on flower-bedecked balconies. A smell of horses and harness and of excited humanity from the bursting crowds pervades the sanded roads. Red buses and lumbering taxis are adorned with Prince of Wales feathers, gaiety apparent on hats and umbrellas, on flags and balloons, on bunting, placards and lamp-stands. And, above, the continual, circling flight of London's pigeons.

Amid all the kaleidoscope, through thronging thousands, the Bride, like a modern Cinderella, comes in a glass coach along the Mall to St. Paul's. And the vast throng surging round the statue of Queen Victoria sways and moves and acclaims as one.

In the strong sunlight, even St. Paul's Cathedral, with Sir Christopher Wren's 'embracing dome' shows golden. On massive white pillars, the rounded mass rises triumphant from the grey pavements, red-carpeted steps and green foliage below. And inside, from the Whispering Gallery, seen through a vista of receding arches, a reflection of jewelled colours is strewn on the black and white chequered floor.

From the mauve mosaics and cupids of the upper arches, the Bride is seen carrying yellow roses, cream gardenias and white orchids, freesias and lilies of the valley. Among trailing ivy leaves are myrtle and veronica, and samphire from the salt marshes of Norfolk is used to garnish the royal breakfast. Nature greatly adorns the wedding, traditional English flowers of garden, woodland meadow and marsh making fragrant this great pageant. And, leaving behind an exhilarated and exhausted London, the new Princess sails away with her Prince into the glory of the sunset skies.

A fit ending to the month of pageantry that is July.

A Riot of Roses

Roses ramble, roses riot,
Roses clamber everywhere,
Bursting forth in gay abandon,
Covering corners long since bare:

Reach up, skywards, in your beauty,
Peach and apricot and flame,
Gay, defiant, shy and glorious,
None of you is quite the same:

Fallen petals thickly cover
Every winding path I tread,
Salmon, orange, white as snowflakes,
Palely pink and scarlet red:

In the twilight richly glowing,
Sparkling brightest in the rain;
In the dew all greenly gleaming;
All my roses bloom again!

Hold high your heads, proud, erect
And all the summer bless,
And fill your world with fragrance,
Conscious in loveliness.

AUGUST

The infant year that was so small, so new, has matured. August, the heart of Summer's pageant, is a month of pulsating life, slumbrous heat and vivid colour; a month drowsy with its own fulfilment; a time of approaching and glowing ripeness. It is a brief period of waiting and brooding.

The trees, resplendent, stand in darkest green shadow, heavy with leaf, rich with swelling fruit and setting seed. Birds, insects and small mammals have reproduced and increased.

In August, 'ripeness is all' and Nature seems to hold its breath and life seems for a while to be suspended.

This year the month comes with heavy rains, swamping the grass and buffeting the trees. Afterwards, my garden glows again, each separate blade of grass gleaming in a rich emerald and every flower, after its beating, lifting its face to a new beauty. The moist soil drinks and drains the welcome moisture and trees spread anew in a Spring-like green.

The first week ends in early mists which lie on the fields like swathes of cotton wool and are caught, entangled, in the hair of tree-tops. Slowly they rise to disclose the moist yellow of buttercups, kingcups and white dewy daisies that star the grass.

Two hot Summer days, torrid with temperatures of 80°F. Cabbage-white butterflies, in erratic flight, appear from nowhere to finally settle with folded wings on cabbage leaves, under which one pale yellow egg is laid singly. Some leaves have already been reduced to skeletons by their devouring caterpillars.

My robin hops near, pausing to peck for unsuspecting insects released by the rain. Always he watches me and always he draws nearer dipping his head as he comes. One of the youngest brood from the nest in the coppice, he has the pinkest of breasts and the brightest of eyes. The whole procedure is as serious and studied as a ritual.

The heat of the day ends with a greenish light on the hills, strange and almost foreboding. Through an eerie gloom that is unnatural, the early evening draws in. My pink and yellow roses lift deeply warm blossoms to glow like lights through the dusk. All life seems sleeping. No hum of insects or twittering of birds presages the night. The world is waiting,

'Quiet as a nun,
Breathless with expectation.'

Afterwards, come violent electric storms lasting the whole morning. As though the very clouds are colliding, thunder rolls around like the growling of the gods. No rain comes and the heavy blue of the skies holds a heat that smoulders. Fields look withered and

hard.

Suddenly, a noon monsoon with beating rains for three hours, leaving roads and vales temporarily flooded. Overcast skies for the rest of the day. This is high Summer! Strange to be bereft in August of all bird-song and movement. Where can all the small creatures be hiding?

A fine day, to begin the second week and wasps abound, with a large nest under the eaves. I wander down Merry Hill to stand in the sun by the five-barred gate. A few quaking grasses survive, of a delicate silver and lavender. There is plenty of sound today. Listen gratefully to the grasshopper's harsh warble and the chirping of crickets deep in the hedge bank. A few years ago, the song of the cricket was seriously believed to presage death.

These rhythmic sounds come from the grass verge which has been left undisturbed. The hot, glancing sun has hatched the eggs that have been laid among the shining blades, where a few remaining golden faces of buttercups gleam.

As I watch, a grasshopper jumps in a flash into the sun's light and near him several orange ladybirds brighten the green. At the Farm the bees have swarmed, the old queen abandoning her throne to a new queen, who has stayed behind in the hive.

The heat seems a concrete thing, producing its own quality of uneasy quietude, and hanging above the parched ground in a thick shimmer of weight that seems immoveable. Though insects hum and drone and chirp, small birds are mute. Perhaps they are momentarily surfeited by the intense joys of Summer.

The near meadow, where several brimstone butterflies have settled on the heads of marguerites, holds a russet haze of sorrel over its harsh green, while the far fields, shorn of their crops, are baked and bare.

Those very crops that were gathered will give life to man. Before, they gave abundant shelter to countless small creatures. Field bindweed now tangles the cut and brittle stems. From them comes a tiny scream which pierces the air. It is probably the cry of a homeless mouse being tracked down by a weasel.

Leaves hang listless in the sultry air and high above, a kestrel hovers, watching with keen eyes a displaced shrew or vole. Seemingly motionless, he can fly at the same speed as the wind, deserving his second name of windhover. The warm, striped plumage of the kestrel is seen more often now over suburban gardens. Although he usually nests in hollow trees, one adventurous bird has chosen for his high home a tower in the City. The female kestrel speaks to each of her fledglings as it breaks through from the shell, encouraging it to make the greatest effort. Mostly content in the Summer with beetles and lizards, these birds feed in cooler weather on small mammals.

My cut and dried meadow is now almost the same hue as the cornfield, but the verges are still beautiful with violet-like blooms of crane's-bill and patches of yellow toadflax. A bumble-bee, in ungainly and clumsy flight, hovers over bird's-foot trefoil and thistle before he settles in the sun. Tiny wild pansies, eloquent and perfect, open their faces offering fragrance, and cream-coloured sprays of meadowsweet are attracting many insects. In medieval days these plants were gathered with rushes to strew on the floor of the house or used to make beds sweet-smelling, in the same way that lavender is used.

August is now advanced and the heat is lifting a little. Swallows fly low after winged insects. It is as though Spring yellow has returned to Merry Hill, for I find that cinquefoil and yarrow are gold and plentiful, while creeping Jenny appears under my feet,

'Quieter than a harvest mouse
Or small raindrops on a house!'

The leaves of the hedgerow, that distinctive feature of our landscape, are still vivid and tiny moths, with faintly mauve edges to their white wings, continually flit and tremble on flower heads. The borage shows in a brilliant blue of star-like blooms. Its grey-green leaves have a cucumber taste and can be used in salad, while its flowers once flavoured claret cup. John Gerard, the sixteenth century herbalist, thought much of borage, declaring, 'Those of our time do make the flour in salads, to exhilarate and make the mind glad.'

The vivid red and poisonous berries of the cuckoo-pint are early this year while umbrellas, cream and heavily scented, of the common elder or Judas tree are dying. These flowers were once used medicinally, while its hollow branches were formed into flutes and pipes. 'Eldern-berrie' wine is made today by many country folk.

A colourful, red-spotted burnet moth, that flies by day, seeks a bloom on which to settle, while a frail peacock butterfly, brilliant of eye, suns itself on a flower of the common nettle. These bright-winged creatures are sometimes called 'shuffle wing' because of their unusual movement. Is this one already thinking of Winter quarters?

Among sparse clover, bumble-bees and flies are busy. The dock is seeded in shades of pink, pale brown and dark red. As I pass I touch the prickles of a rose burr or Robin's pincushion. It is actually a gall caused by the gall wasp, which lays its caterpillars inside the burr, where they spend the Winter. On one curling, russet dock leaf a green-veined white butterfly has settled, while, high above, the white rumps of house martins show as they pass.

The middle of the month and the continued heat calls me away to the 'coolth' of East Anglia; to the soaring trees, the tranquillity and 'interminable skies' of Constable. Alas, the Vale has lost many elms and the sylvan solitude is threatened both by tourism and modern farming methods.

Beyond rough pastures, where companionable trees give shade, placid cows stand cooling in the shallows of the river. Overhead, in the still afternoon air, a lark is trilling in pure, unadulterated joy. The heat intensifies, the only movement being the endless dipping and diving of swifts for the yellow flies they love.

An occasional moon-daisy or blue cornflower relieves the hard green of coarse grass. Over the hill's crest, where the 'wheat-filled earth', flattened by the recent rains, is filling, mayweed grows. And, between the wheat's rows, yellow charlock lifts and 'corn poppies that in crimson dwell'. On the poisonous ragwort, a tiger moth has laid its eggs and now its caterpillars swarm among the leaves.

The sky is like liquid glaze, blue and limpid, and even the hidden insects are languourous. Gratefully, I linger where the hazels and alders form 'a tunnel of green gloom' and 'the stream mysterious glides beneath'. The tree-tops move with a gentle murmur, as drops, clear as crystal, fall softly over sparkling stones. A cormorant which has moved up river from the estuary, stands motionless, its wide wings outstretched over the water, looking for food. They seem to travel far afield. Recently, one roosted regularly on the Houses of Parliament.

Around an oak tree a purple emperor flits. Its eggs have been laid on sallow leaves. Cautiously, I draw near, when he settles. This one is a male one since it has imperial purple on its wings. In the thorns of the bank, where the reeds lean tall, a willow-wren has hidden her nest, carefully forming it with a grass roof and a feathered floor for her tiniest of babies.

A few green leaves from the white-barked alder float lazily with the water's current. Tall plants are flourishing. Purple loosestrife, thistle, rosebay willow herb, belled campanula of mauve and white, spiky rushes with pink flowers and golden florets of agrimony, are high and water-leaning. Before I come to it, I smell the delicious scent of flowering mint.

On the grass of the river's margin a caddis fly has settled and in the bank, the unseen badger has successfully built his chamber with ventilating shafts, where he can live as long as fifteen years. It is good to hear that the gassing of badger sets has ceased.

Here, near the path, the remains of an eel lie half concealed, the work of an otter, a heron or even a mink, since in the remote regions, the latter have increased after escaping from a mink farm. Smaller than the otter, the mink can roll and twist at incredible speed, throwing up the water around him in high fountains. In another part of the country, animal lovers have released many mink, which have, unfortunately, done much harm to the surrounding wildlife.

The quiet splash of a water-vole cuts through the air and he is gone, leaving receding ripples on the surface where he has been. This small animal prefers to feed at dusk, mostly on grass. His plopping noise is an oft-heard sound on the river. I do not see the pike-like zander, which escaped from captivity years ago and is rather vicious.

Round a bend in the river I come upon a profusion of water forget-me-nots, glancing blue-eyed. Great water dock and yellow iris lift dying heads. The large dock leaves hold slugs, small and horned.

The quick, blue flash of a kingfisher comes and goes, that bird that is never far from water. He is a fair-weather bird, however, for in bad weather he will soon move away and take shelter. Being an excellent diver, he catches his fish and bangs it on a stone before eating.

Glittering in sun-splashed yellows and blues, a dragon-fly hangs on the air for a second before it vanishes. Where water mint and the white, leafless flowers of water buttercup grow, a family of moorhens make their quiet, shimmering way among the reeds.

Hoverflies and black beetles are interested in the paling cups of marsh marigold and little silver fish, small fry, leap through the air for flies.

The distinctive cry of the hoopoe comes to me across the waters. He is a fine bird with a salmon and black plumage, a crested head and long, slender beak. He enjoys grasshoppers, but tiny frogs and small lizards are dashed on a stone before being eaten. He is not visible, for he is well camouflaged and has been here since April, but will soon be leaving for Africa. In tree hollows, four grey-white eggs have been laid, the young being fed by the male bird with beetle larvae.

Quietly, I wait and watch while, at my feet, a baby sand-lizard emerges from a sand hole where eggs have been laid in June. It is a brilliant gold and flicks its tongue incessantly for food, looking for heather whose nectar it loves. Quickly, it slips away and disappears.

A solitary loneliness broods over the coast, where it is easy to imagine the forlorn bells of Dunwich village, especially when across the wastelands comes the wild and mournful trilling of the stone curlew. Listening I wait, as submerged forever beneath the waves, the bells seem to be joining his sad song, 'Tolling for me and my sole self'. Or could the sound come from the myriads of heather bells?

To wander round the Heath is rewarding, especially as this type of land is becoming a rarity. One of the oldest of landscapes, the Heath was cut for fuel, being a clearing of ancient man. Among the sharp, thin grass, I find broom and heather making a riot of purple and yellow. Many spiders scuttle under the broom. A few bushes of rich gorse are still in flower and here, among the dunes, the cinnabar moth, flying by day, lays its yellow eggs. Its caterpillars feed on the plentiful groundsel and coltsfoot or even ragwort, though birds shun this poisonous plant.

Through the silence comes the sound as of two pebbles being rubbed together. That will be the stone-chat's song. Not far away both wheatears and stone-chats nest and breed on the sandy ground. Here I find the thick, edible leaves of the button-hole plant, whose small, white flowers, that grew from the leaf's centre in April, are now dead, leaving behind them a big black seed.

This is the haunt of the voracious 'butcher bird', the great grey shrike, where he preys on baby finches, flies and bees. He is hidden, but the bones of his captives hang on thorns as evidence of his passing. I am told that the snow bunting sometimes breaks his migratory flight here in order to feed and rest.

On the edge of a little wood, rust-spot fungus of coffee and cream is growing and, under an oak, the pink tinge of russula vesca shows among dead leaves. A large heath butterfly rests on a fine show of heather and, on the bark of a tree, I see an emperor moth, immobile. The overlapping leaves of ling grows near, which retain water and are liked by the caterpillars of this moth.

Where bees are busy among the bramble, flowers and fluffy balls of thistle float, and I find tufts of little cluster fungus, orange and cream, covering the base of a tree stump.

Around the Washes of the Ouse there is cause for rejoicing. The black-tailed godwit, which forsook them, has returned after an absence of a century. In the creeks here, washed by sea and wind, tiny finches crowd around flowers of the sea aster, and shelduck and redshanks feed on the edge of the water.

The cry of wild birds and the eerie sound of the wind in the thick swaying reeds give the neighbouring marshes a strange sense of unreality. Here, the great crested grebe, with her striking orange and black ruff, sits on her floating nest in the swamp where she will rear her chicks. These birds have an elaborate court-ship ritual, confronting each other with outstretched, nodding heads and presenting each other with water-weed and building material for their raft home. It is delightful to see the young chicks riding on the backs of their parents. These young have white stripes and are often given small feathers to help in digestion. Thought to be extinct in the nineteenth century, the great crested grebe is now happily resettled.

Among the swamps the coypu lives. Imported from the Argentine for fur farming, these brownish animals still flourish in the wilds of East Anglia, causing much damage to farm and reed crops, where they burrow under river banks and even invade gardens.

From the midst of the swamp I hear the deep boom of the elusive bittern, but do not see him. He has a heavy flight and fierce dark eyes and, with a straight and upturned head, sways from side to side on his neat nest among the reeds. If disturbed, he stays perfectly still, head in air. Soon he will migrate to Africa.

The frail, scarce, darter dragon-fly flits by. These creatures have bred here in June, the male now being powder blue while the female has tawny wings. Down in the ditch lives the great silver water-beetle, our largest, which often flies at night. A rare sight is the natterjack toad which breeds in the fresh pools. Very soon now he will be hibernating in the dunes.

The water-boatmen are busy. They flourish, preying on the hoverfly. Climbing on top, the water-boatman drags his victim below the surface of the water before eating him.

Perhaps the most beautiful creature here is the rare and graceful swallowtail butterfly, moving delicately over the marshes of Norfolk and mirroring its image in the pools. Not far away, the uncommon bee orchid has been found in one of only ten sites left in England.

My wanderings are rewarded by a sight of wonder near the coast. Stretching away to the horizon, are misty fields of purple and gold, for side by side with the scented lavender is the yellow gold of ripened corn. As I draw near, I see that one field has patches of marigold among darkest purple and palest pink. Years ago this sweet-smelling lavender was harvested by women with sickles. What a spectacle for a poet or an artist!

Within sight of the sea, growing on the flat lands that are in Winter constantly under water, sea lavender stretches in long and misty vistas, rivalling the seas themselves in loveliness.

Homing in the last week of August, the landscape seems to be changing from its age-old hues. England's green land is still vastly pleasant, but among it, vivid yellows and blues are appearing. I see that one experimental farmer is growing blue, yellow and white lupins, the pods being protein-filled for animals, while brilliant gold and pale blue comes from oilseed rape and linseed.

My garden welcomes me, seeming richer than ever. A small tortoiseshell, brilliantly beautiful, spreads its wings to the sun. A cluster of them are on the pink valerian, which is still prolific though faded. The delicate green of teasels, their cone-shaped heads flushed with rose, stand erect and tall, a perfect foil to the mauve and white bells of campanula. Crumbling sculptures, like fallen Roman emperors, are crowned with green-leaved, trailing stems of morning glory.

The long spell of fine weather continues and a small, long-tailed spotted flycatcher darts, arrowlike, from the eaves for flies. He doubtless nests in the coppice. I hope I shall see him again. Syringa scents the air, near the tiny pink and blue buds which are swelling on the Michaelmas daisies. I watch two large white butterflies chase each other, before finally settling on cream-flowered privet.

A greenfinch has descended on a dying sunflower, where seeds are beginning to form. With his pale green feathers and blue and yellow wings, he is a handsome bird. I must search in the yew of the coppice, where he probably lives.

My blackbird is unearthing semi-mature lettuces, in order to get at the insects beneath and slugs have annihilated other lettuces. Some cabbage leaves are reduced to small holes and one upstanding rib.

During my absence, marigolds have erupted brilliantly in vivid oranges and yellows. Surely they are a flower of beauty and useful too, being treasured as a medicine since the twelfth century.

The last week and, between moonset and sunrise, a little rain falls, ending the intense heat. On the Hill, chickweed, groundsel and nettles are reflowering. A red admiral, in striking colour, hovers over the garden flowers. It is believed that this butterfly, like birds, has his own territory, from which he discourages all other male red admirals.

Sadly, the large, crumpled, silken petals of Oriental poppies have fallen, leaving deep-cut foliage, finely furred, and silver-haired stems. The big heads are seeding. Some of these seeds are still used to sprinkle on cakes and bread.

Two bullfinches visit the forsythia bush, its flowers long fallen. Surely their plumage is now less bright?

August is ending and, after many days with temperatures in the eighties, comes a morning mist which persists all day until dispelled by a light evening breeze, which rustles in song among the heavy-leafed coppice.

Suddenly, the garden looks decidedly autumnal, as though the pageantry of August had never been. Leaves of hydrangeas are turning rust, with large, drying flower-heads of green and rose. The magnolia's foliage is now in subtle shades of maize and gold. A solitary cabbage white butterfly flicks and flutters as though lost, through the still air. And two toads have emerged from the compost heap, intertwined. They have evidently mated and will go around in close embrace until their eggs have fertilised.

Above the mustard yellow of charlock, an orange-tip butterfly flits. He also loves the lady-smocks of Merry Hill. The beautiful comma butterfly is unseen. Rare, even in Hertfordshire, he likes the waxen-sweet escallonias and visits leaves to drink their raindrops.

Dainty pansies are in seed. Wine made from these flowers was said to be delightful and once, a cure for fever was made from crushed 'pansie seedes chopt verrie fine with a bigge spoon of lad's love'.

Evening. The gnats begin to gather and a white tiger moth flits by. The giant sunflowers that have dominated the garden seem to hold all the gathered heat of August. Bold and beautiful, they tower above all other blooms, greeting me with wide faces, yellow petals still open to the last rays of the sun. Gazing into one mysterious centre, I try to trace the whorls of seeds in its heart. Complicated and complete in design, it is a maze of ingenuity, a small world, part of the vastness of infinity. Its head slowly droops; as though with the passing of the sun's power, its own brief beauty will soon be gone.

A calm lies over everything, as though the pulsating life of thicket, hedgerow, bush and tree were suspended. As the warm darkness deepens, scents become more subtle, flower personalities more apparent. Hollyhocks roundly gleam; marguerites lift moon-faces through the shade; small, velvet pansies are eloquent and wide-eyed; roses glow and elegant Madonna lilies hold graceful heads, queenly and pure, to the setting sun.

On Merry Hill all is subdued by rest, all sound and song is silenced, all fragrance diffused. And, from another hill, in another August not long ago, I remember briefly when the darkened heavens were imprinted with the colour and imagery of fireworks; when softest flame was swiftly transmuted to gold and rose and sea-green. When, through the rich darkness, in cascades of light, flung flowers opened in radiance. And all England remembered with gratitude our beloved Queen Mother, as they watched the pageant of the skies.

Among the trees of the coppice, birds sing their anthem to the dying day, each from his roosting place through the cool and fragrant air; the chaffinch, the yellowhammer, the song-thrush, the blackbird. Even the magpie and rook add their own sounds to the pleasant cooing of lazy wood pigeons. More and more, in the darkling light, the swallows gather and lift away on their journey to the southern sun, higher and higher, like a long trail of pale smoke in the sky. My wrens chirp softly, sleepily, and across the Hill, a nightingale pours out his song in notes that ripple like water falling from high hills.

'This is the pleasant time
The cool, the silent, save when silence yields
To the night-warbling birds.'

Stars now begin to flicker and the whole of heaven seems to stir and tremble. And, suddenly, there is a sadness to this last day, as though not only August, but the whole year, has melted away, taking with it its own perfect and peculiar beauty.

Reflections on a Sunflower

Compact, complete, & darkly green
An inward eye, in depth serene ;
A mini-world, close held, is here,
Cushioned with order, void of fear;
Brown as the earth from whence it
sprung, in secret, burning beauty
                              hung.

Absorbed, I watch and wonder, gaze
Entranced, & lost within the maze
Of myriad spheres of amber, gold:
of whorls & whirls that tightly fold
In coil of gold; of countless worlds
That each within its orbit curls.

Bewitched and drugged as in a dream
  My thoughts fly round within the
stream of endless beauty, inner light-
And from this haven of delight,
Radiant petals, warm, unfurl,
  Outfling their golden tips to curl.

Burnished with light, till day is done,
  They borrow brazen beauty from the sun,
Till, weaving their wondrous way to peace
  I stand renewed, in sheer release :
A paean of mighty music roars,
  Cold earth is lost and heaven soars-
Such tiny worlds of loveliness
              transcend Earth's gloom.
    And shall this promise end?

SEPTEMBER

*'The rusting harvest hedgerow
Still the Traveller's Joy entwines.'*

August has slipped into September which is full of a golden stillness. Yellow-greens of Spring and blue-greens of Summer are mellowing now to a soft ochre or deepening to a tapestry of dusty pinks, wines and bronze. No stirring yet of breeze or rushing of waters and, from birds, low and sweeter songs. The month brings many precious days when Spring, without the urgency of birth, seems reborn.

Three quiet, dull days usher in this September. Groups of colourful tortoiseshell butterflies are a constant source of pleasure among the late valerian flowers and bees are hurriedly busy on seeding sunflowers, as though they sense that the time of their gathering is short. I find them on the hibiscus, too, where a delicate light shows through the petals of violet and blue. Dahlias glow with richness and my Japanese maple is warm-leaved. Hydrangea blooms are dying in green and pink, but are still visited by large white butterflies. Among its yellow and heart-shaped leaves, convolvulus berries already show small and orange, while spindle leaves have a delicate Autumn beauty.

On the apple tree, teasel heads and white fruits of snowberry, blue tits are feeding. These little birds play a macabre game with the bees. Waiting until one returns to its hive, they pounce and enjoy its load of honey. My Japanese anemones have waited until the end of Summer to grace the garden with their wide and softly pink flowers, among frilled leaves. They have almost a quality of Spring.

The rainless week ends with a day of perfect Autumn radiance and my blackbird sings from the chimney breast.

The butterflies this September are wonderful. A peacock, its eyes brilliant, flutters across to cling to a nettle flower. They seem so leisurely; the flapping red admiral, the easy-going large white, the lazy peacock. Yet, they can fly faster than man can run and can move at twenty-five miles an hour during migration.

The excessive dryness is bringing down a few early leaves from the beech of the coppice, where the horse-chestnut and sycamore are already paling in shades of corn and bronze. Honesty is silvering quickly and roses are beginning to fade.

A blowfly invades the house; the first wasp appears, buzzing energetically and enjoying the juice of blackberries. When he has gone, a red admiral takes his place among the ripe fruit.

Raspberries are fruiting well. Centuries ago, this luscious fruit was grown on the slopes of Mount Ida, but our cultivated ones originate from wild canes nourished by Scottish crofters.

Small white butterflies rest briefly on the remnants of sunflowers and my blackbird seems busy about the compost heap. Has he had a fresh brood there, I wonder? Sweet peas are blooming anew and every evening brings its whirling mass of gnats.

For the first time in three weeks, a drizzle of rain, followed by a sudden, unexpected, capricious breeze, which rustles through the limes of my little avenue, where the leaves, once so vivid, are now pitted and harsh. Underneath, I find single plants of clustered bell-flower, hairy willow-herb and common rock-rose. Near by, the willow shows a hint of blue, where it sweeps and weeps in the wind.

Among the grey-green foliage of the coppice, the turmoil tears down leaves and strews twigs on the ground.

But Summer has by no means gone. Another glorious day, with breezes delicately flower-laden and a luxuriant air in the garden. Over a mallow flower, which he probably knew as a caterpillar, a painted lady butterfly hovers in the warmth. Rosy valerian persists abundantly; giant sunflowers are splendid and regal, with golden crowns on their heads and the bay tree shows purple-black fruits. Marguerites are fading in the drought, but many rose-buds retain a delicate beauty. I discover several yellow poppies, a late mauve campanula, asters, phlox and petunias, while the golden blooms of Rose of Sharon, along with golden rod, still glow.

In the night, a sudden shower saves many more flowers and morning brings my thrush and a bright chaffinch to probe for worms on the lawn. Only fifty years ago these birds were used as baits for hawks, after being kept for a long time without food and water.

The earth freshens and breathes again; spikes of mauve appear on the mint and my dwarf cypress is saffron against the dark green of drying rose leaves. A garden tiger moth, with orange on its lower wings, hides under a leaf before hibernating.

Mid-month and the sky a colder blue. A few desultory showers and evening comes early, bringing a strange light on the horizon with ominous clouds of violet and indigo above. And down comes the rain.

Afterwards, Merry Hill has a new, subtle and mellowing beauty. In the hedges, threads of convolvulus twist and from the soil a tiger beetle emerges, with formidable jaws ready to catch his prey. The ribwort is seeding in my meadow and

'How sweetly smells the honeysuckle
In the hushed fields!'

I watch the erratic flight of two small white butterflies in their dance of courtship near a wild cabbage leaf, on which they produce a third generation.

An evening stroll in the dusk of Merry Hill is rewarding, but where have all the glow-worms gone and with them the tiny light from the male beetle, with which he attracted the female? Once, they were used indoors as a light and twinkled in the hedgerow like little, glancing stars. Wild parsley still borders the lane, with traveller's joy, and near the coppice, wood sorrel lifts dainty flowers, so soon to die among clover-like leaves.

A flan can be made from the ever-flourishing dead nettle leaves, by adding milk before cooking. This plant was once known as Adam and Eve in a Bower.

In the undergrowth of the ditch, a tiny shrew mothers her litter of tinier babies. Among the oak leaves grasshoppers chirp loudly in the evening air. On the faded yellow of common rock rose a brown argus butterfly, almost invisible, has settled, its wings orange-flushed. Is it thinking of hibernation?

At night there is much activity under the protective layers of dead leaves, here in the hedge. Spiders, which have taken a whole year to mature, are busy and male bush crickets sing, their long, hind legs enabling them to jump long distances. The hedge beetle is using his huge jaws for his vicious attacks on his prey.

Round a curve of the Hill, on the next morning, I come to a scene which is like a miniature country pageant. At the entrance to my meadow, by the five-barred gate, a farrier tries his skill on a child's piebald pony. Patient pony and anxious child, with reassuring hand on bridle, wait while the shoe is adjusted. The craft of the farrier is two thousand years old, the Romans being the first to use iron shoes on their cavalry horses.

The meadow, dry now, where the white clover is browning, has stunted plants struggling to produce new shoots. I catch a glimpse of a fox under the far hedge. There are several families of foxes on the Hill and their tiny babies, born blind, are now playful cubs. These young are not about, as they hide in the daytime in rabbit burrows or among dying bracken and fern.

In the hedge

'Up tall
Turrets of sorrel, the bindweed climbs,'

still flaunting a few white flowers. Here, too, a holly blue butterfly, one of a second brood, rests on ivy buds.

It is cooler in the third week, leaving behind me as I go north, many devastated elm trees. What a beautiful thing a tree is and how sad to lose them. Small wonder that our remote ancestors worshipped them.

An ancient tree is a pageant in itself. Many live longer than man or any other living thing, rivalling the age of elephants and giant turtles. It has been found, by the discovery of pollen, preserved in peat, that birch trees lived ten thousand years ago. One sequoia tree in California, is reputed to be the largest living thing in the world, while in Queensland the macrozamia is believed to be twelve thousand years old. Giant redwoods may grow for three thousand years and many olive trees, which shed and renew their leaves continuously, are as old as Christianity.

The dragon trees of Teneriffe are mighty, where, in the not so distant past, a chief would entertain his friends but never his enemies. The dour yew with berries bright as blood is long-living. One mighty yew hedge I know, which spans a rectory garden, has known four hundred summers. An old spar has been unearthed that is made of yew that grew two hundred and fifty thousand years ago.

In increasing numbers, I pass the blighted elms, shorn of their nobility and grown skeletal. For the past fifty years they have been attacked by the ambrosia beetle, which is stemming the flow of the sap. No longer can one sight the graceful outline of elms crowning a hill against the sky and a unique scene of English landscape has gone forever. Like the witches of Endor they stand now, incredibly sad with dying arms clutching the air. Happily, an injection has been discovered that will temporarily arrest the disease.

In Britain, which has every variety of tree from most other countries, the extreme weather has this year killed a number of cedar, cypress and eucaluptus, along with many shrubs. The sap has been frozen in the trunks and roots and buds have been destroyed by frost.

Embedded deep in its own earth, a tree gives so much to man; beauty, usefulness, shade, a home for wildlife, a feeling of solidarity, of permanence. Yet, within the last forty years, half of England's ancient woodlands have disappeared.

Leaving far behind me the ravaged trees, I move across a wide heath, surely a 'heather-honied hill', one of these clearings of early man. For Dorothy Wordsworth these regions 'called home the heart to quietness'.

The calm hills lift to a blue sky, flecked with passing cloud. Below, juniper clothes the crags, where a hawk wings from his eyric. The lofty solitude, capped by peaks and rocks, is a world of shadows, with remote slopes and black ravines.

Here, on the lower hills, the bracken is withering in flames of gold, the furze dying.

Among the tufted grass I find a few speedwells lingering, along with vetch and tormentil. Beyond, in a recently cropped and ploughed field, two hares run frantically in a crazy effort to find a fresh home for themselves and their depleted family. In the face of danger from a larger animal such as a dog, they would become immobile, but now, terrified, they flee blindly before this unknown peril. Above them, crows caw and cry, black wings circling round.

Much of the heather is now dry and brittle, but grouse are fattening on the new shoots opening to the sun. He can out-flight the falcon. Ling grows here and bilberries are ripening. Many heather bells still swing in the breeze and welcome the bees. These flowers are much liked by emperor moths, too, that fly by day in a zigzag fashion and have bred their caterpillars here.

I move cautiously and am rewarded, for near me a skylark lifts in the air in a spiral of song. Surely he is the last of the year, for soon he must join his flock in migration until next February. Among the mounds of rich, lilac heather shining in the sun, moths, bumble-bees and butterflies feed on the nectar. Briefly, a superb pheasant shows his handsome plumage. Originally from Asia, these birds were kept in pens by the Normans and much appreciated.

In little damp pools tiny sundew plants thrive, ready with sticky traps for unsuspecting insects. The yellow flowers of bog asphodel are dying but its stems are warmly red. Further away, a few stones among sandy ground harbour a smooth snake and an unwary lizard. The latter must be careful or the snake will swallow him head first, as well as many of the tiny creatures that abound here. The nightjar is only a Summer visitor and may have flown. He has a long tail and is carefully camouflaged, seeking insects on the wing.

Growing from rocks, pink stonecrop with its attendant butterflies is plentiful. The drystone walls have hanging cascades of miniature ivy and tiny herb Robert, with here and there a rare yellow poppy and harebell, both of which seem unaware that it is Autumn. As I watch, a plump partridge bustles clumsily out of the undergrowth near and raises heavy wings, disappearing in an awkward flight.

Into the lake below, a beck flows gently with a sad, sweet sound. Many jackdaws fly by the water, never moving far, as Winter approaches, from the warmth of a cottage chimney. Flies still rise and drift over the surface, where reeds grow green again and willow-herb rosy, in their watery shadows.

At last I come to my hanging wood, which still holds the warmth of Summer with glades of 'beechen green and shadows numberless'.

But the wood has changed. The greens are mellowing, the rowan is berry-bunched, brambles and rose

leaves are reddening and the birch trees wear their Autumn dress of soft yellows, while maple trees are aflame.

Pink campion shows in delicate drifts and rose bay willow-herb has invaded the glades, while on sturdy stalks wild arum has orange berries.

Yet the wood still has a softened beauty and its own sounds. 'In a forest are many voices; a man must hear with his own ears, the carol that is for him'. Yellowing oaks spread wide above lemon of rock-rose, gold of cinquefoil, violet of scabious and yellow of snap-dragon. And the fragrance of thyme covers all.

The rare pine marten lives here, but I do not see his chocolate fur and cream and white throat. He is much too wary. Until a hundred years ago he ravaged poultry and ewes mercilessly. The few remaining ones have now retreated mainly to the larger forests where they hunt the red squirrel.

On the outskirts of the wood, I have a good view of another cock pheasant in gold, red and green feathers and trailing his long and splendid tail as he feeds. A slight movement from me and he runs and scuttles quickly over an open space to shelter.

I pause to watch a red admiral settle on a late scabious, as the call of a willow tit comes from the trees.

Heavy rain is followed by wind, but the twenty-third of the month brings brilliant, glittering sunshine. I move on, for ahead of me are 'those blue, remembered hills', with all the pageantry of Lakeland. And beyond them, in splendid isolation, the Pikes rear. Against their darkness a shaft of sun lights up the soaring wings of a golden eagle.

The Fells, paved with lichen and ice-age rocks, have a quality of eternity. Here, the bracken which only last month was a sea of green, has turned gold and bronze. The quietness is broken only by the age-old song of the waters, falling down rocky hollows and around stony outcrops to the valley below.

The good weather holds, sun soon dispelling the morning mist. But a dampness rises from grassy tussocks, peaty earth, vivid moss and glistening rocks. Numerous becks and forces tumble through the bare hillside. Many herons keep silent vigil by the waters, while the buzzard, on wide-spread wings sails on patrol above. The silence is so complete that I find myself listening to its unheard music.

Autumn

Corn-filled earth the trees enfold:
Poppies red and
Fields of gold :
Toadstools lift
And chanterelle:

A lone and lovely flower bell :

Bronzèd beech and
Sunlight mellow
Wrens all gathering
On the willow :

Alder, elder, aspens weeping:
Scene of rest and
Time of sleeping.

By an upper tarn, dark and mysterious, a curlew with down-curling beak, probes for worms and insects under the water. In these upper regions, I hope to see a rare mountain ringlet, that bright butterfly that loves the heights. It breeds in small, deep hollows near water, flies low in sunshine and has laid its eggs on mat or rush grass. It may already have settled in hibernation deep in some yellowing tussock.

Between rocky ledges and hanging valleys, the Herdwick wander, grandmother, mother and son together. Very hardy, these sheep can survive in snowdrifts for two weeks, eating their own wool and having a strong homing instinct. They have one main enemy, the fox, which in these parts is big and ruthless, with a liking for weak lambs. The Lakeland shepherd, however, is wise; knowing his own foxes, he hunts them on foot.

Sudden rumblings. And the Pikes, a formidable freize in purple and indigo, echo back the thunder's greeting. Black clouds now shroud the hills in an ominous haze. Swift squalls of rain and the whole scene changes. Everything grows fluid, awash in shining liquid. Numerous silver ribbons of water steam and tumble, and impatient becks plunge around the rocks to drop through the bracken of burnished gold. And, above, angry vapour floats from one ridge to another.

Through the darkening skies, a kestrel soars with spreading wings over lake and cliff and

'A raven croaks from a craggy stone,
In the black of the wind he croaks alone.'

From a safe distance, rooks call. They may drive away the kestrel or a hawk, but not the savage raven. Even the buzzard fears him.

The storm ceases and the imperturbable sheep graze on, Byron's 'wild flock that never need a fold'.

The month is ending but it is quiet down here again in the valley. The lake is rippleless with a floor of glass reflecting a passing lapwing. After its sudden surgings, the river meanders at her own sweet will, taking a wide flowing through meadows. And, round a bend and standing in still waters, a solitary heron waits.

In the sun's light, brown trout snatch at the remaining flies where saffron rushes and dying reeds go down to the water's edge. Among the whitebeam and wild cherry, a few yellow flags still blow, near the darkening blue of bugle flowers. By a peaty pool, pink of pimpernel, blue of harebell and yellow stars of asphodel, show scarce blooms.

Flecked with green lichen, drystone walls climb upwards to the fell. Many pied wagtails run and feed near; their young have been raised here in their nests built in the holes of the walls. They like the reed beds and will soon be gathering in large roosts for the Autumn. Under a hazel tree, a thrifty red squirrel has strewn the ground with nibbled nuts. And, towering over all, the Pikes point, impressive, immovable.

Over the peaks, white clouds pass slowly, softening their contours. The drifting mists lend an added charm to the hills, distilling their beauty. Now, the topmost crags are hidden, shrouded in 'this blended holiness of sea and sky'. Soon, they will be 'clothed with the heavens and crowned with the stars'. A shepherd and his dog seem frozen into this wild and craggy landscape and the wings of a high-floating windhover seem motionless.

The sky has the silken sheen of the sea, with puffs of pink and gold cloud. And, sudden, the last, clear light of the sun gleams on hills transparent, serene, while, beyond the Pikes, the horizon is suffused with a luminous green and the little clouds form into ships riding an ocean of blue majestically, their sails tall and widely billowing.

And, in a trail of veiled loveliness, and the fluid flight of wild geese, September is gone.

TARN HOWS

See, where the
       deepening tapestry unfolds
Here as the
      steepening mount we breast,
Till our delighted gaze can rest
      On this wide bowl of beauty:
Surely the peace of angels' wings
Hovers around this hallowed place :
Below, and far, the waters of the lake
Hold each tree-crowned isle
        as in a diadem ,
And pine-woods rise, & tawny clouds
Of turning leaf in sculpted mould
Lie in enfolded green, soft and serene
As the blue drifts high —
      And all the world is one ,
while the very heart of me
      Absorbs this whole tranquillity,
Till, from the lake's still deeps,
      the bittern calls and cries
      and soars above.

OCTOBER

'Alone upon her starry way,
The moon to fullness grown
Moves, shining, through her misty meads.'

The month of October presents one of the most colourful of the year's pageants. It is a time of fulfilment, since Autumn is synonymous with harvest. While September holds mystery, October's gifts are of the earth, earthy.

There is little mystery about October. Her tapestry is bright with the splendour of flamboyance in fleshy fruits and full gourds; plump fish and well-nourished fowl, ripe seeds and rich harvest.

Although October's voice is boisterous, its few tranquil days of golden stillness can deceive one into imagining it is Spring again, in the last defiant gesture before the onslaught and austerity of Winter.

When light mists gather above the Hill, the first few mornings of this October bring a stealing chill. On the horizon smoke-plumes lift from the earth that is being fired and purged for ploughing. From the cleansed soil, a hare suddenly races for cover. These animals, so delightful to watch, playful and swift and slightly mad, are growing quite rare.

Tawny wheat, golden oats and full-bearded barley silvering in the sun, all have gone, leaving a pink-bronze stubble, where flocks of finches, yellowhammers and starlings search to split the fallen grain. Insects have been unearthed or are being sought among the few harvest weeds that still remain.

On Merry Hill, the plough turning the earth for next year's crop, has many followers. Strong-winged gulls, jackdaws and peewits weave around for beetles and

'With cautious side-long looks,
  Are following, close behind, the rooks.'

The latter, looking portly as old gentlemen, seek the earthworms thrown out by the upturned soil. Flocks of bramblings have come this month from Scandinavia to forage, although they prefer beech nuts.

A young, straying pheasant, however, will soon be stalked by the fierce weasel who watches unseen. Tiny, red-haired and immensely agile, this creature stands up on his hind legs, sniffs the air with suspicion, and 'pop goes the weasel'. Swift as a swallow, he can strike. In the depths of Winter, when he is desperate for food, he may come near the farm where he is useful, since he preys on mice, voles and young rats.

A sudden scuffle across the brown earth. It is a rabbit, but all I see is his white scut. As he is so vulnerable he has 'gone to earth' but, unlike the foxes of Merry Hill, he rarely escapes his predators. Introduced by the Normans and used since then for meat and fur, the rabbit is responsible for ruining the bark of many trees and his close grazing undoubtedly alters the contours of our landscape.

Two lovely, balmy days in the second week, with temperatures of 60°. Merry Hill is bright and glows with well-being. Paled in the sunlight the beech, rivalling its splendour of early Spring, spreads in a golden bronze. A tree that has always been beloved, it was placed under protection in the time of Elizabeth I. Fingers of horse-chestnut are saffron, oak leaves are tinted in soft tones of ochre and russet, and birch trees, thinning, are a pale yellow. Hazel nuts and sloes, walnuts and chestnuts are all plentiful this year, while large bunches of elderberries glisten, most attractive to the willow warbler with his laughing note. But, alas, most of these charming birds are departing or already gone.

I pass the old barn which is now growing empty of hay, but not of owls. I wait, but not one is visible, as he hunts in the dusk, having the useful habit of being able to swivel his head in a remarkable way.

'That other delightful owl,
The short-eared owl'

was once a cave-dweller, but now likes to be near man and his food of mice, sparrows and shrews. The tawny owl prefers Whippendell Woods, hiding himself in dense foliage and building his nest in a hollow tree, from which he swoops, holding tightly with his talons and often swallowing his prey whole.

There seem to be a reblooming in the hedgerows, where the Summer-like sun intensifies the rich colours. Hips hang in a crowd and are crimson; guelder rose shows darkly red; berries of black bryony sparkle. The beautiful but poisonous black fruits of deadly nightshade still have a few bell-flowers around them. Wide arcs holding juicy blackberries spread out to the sun and, clinging to their stems, draggled white petals still remain. Pips of this fruit have been found which are as old as Neolithic man.

Sloes are ripening and the unusual-shaped fruits of the spindle tree are orange and purple. Oak apples grow and the brier holds Robin's pincushions. The gin that can be made from sloes is smooth, full of flavour and simple of recipe. Where the wild rose grew, hips are now vividly orange. Creepers of dying goose-grass wind around stronger branches and tiny bitter apples are forming on the wild crab. And, everywhere, convolvulous clings with a tight grip to stems and fences.

Wines can be made from many flowers and fruits, the flower wines being quicker to ferment.

Frail skeletons of cow parsley border the road by the browning nuts on the hazel. I taste one of the nuts, but they are not yet ripe. Since time immemorial, the hazel has been connected with poetry and wisdom and its fruits regarded as a nutritious food.

Among these plants, a mouse, driven from his home in the field, is sheltering. If this hedgerow is laid low, he will be bereft.

My meadow still holds a few belated flowers. Common toadflax is yellow and orange; a scarlet poppy blows; one campion grows in isolation, but rose bay willow-herb is rampant. Where I walk, a tuft of light red berries rises from a plant of lords-and-ladies.

Ahead of me, two small heath butterflies dance and flit above rough grass heads. Dainty and warm-coloured, with a prominant 'eye' on their forewings, they are a second brood, twice welcome. Their 'eyes' are a protective device to discourage insect-eating birds.

Everything, everywhere, in these precious days, seems to glow with reflected light. And, over all, the spiders climb and weave their frail, gossamer webs, where each separate strand silvers in the sun. These webs are strong enough, however, to trap and hold grasshoppers, bluebottles and beetles. Interesting to read that some spiders are enterprising and take flies caught in other spiders' webs.

A close cousin, the house spider, can be found through the year indoors, where he can live in dense corner webs for several years.

It is mid-month and, after a damp and cold day, the hills are shrouded in early mist which soon rises to disclose small groups of wild mushrooms, wet with dew.

Over the coppice rises the pleasant smell of burning, Autumn leaves. A wood pigeon flies long before dark to roost in my sycamore. Perhaps he objects to the raucous cry of the magpies. A sudden wind creaks the branches and the trees bow before it, lifting white undersides of leaves. On the birch it quivers the hanging, heart-shaped leaves of molten gold, whirling in little flurries those already fallen. But, though yellowed, some are still reluctant to fall. Apart from the leafless privet, the trees and bushes are wonderful this year, retaining thick crowds of foliage. There is little sign, as yet, of them falling *en masse*.

The first morning frost heralds light and sunny weather, full of quietness in a drowse of Autumn. But soon a mist of smoke clouds the tree-tops and, from many bonfires, a delicious smell of burning pervades the air. Solitary leaves spiral noiselessly down and the atmosphere holds a sense of repose. Evening comes quickly.

In my little avenue, the limes wave wildly, frantically, as though trying to free themselves. Large pointed leaves cling loosely but there is still a remnant of growth below them. Intermittent sunlight dapples the path, showing pale pink parasols of yarrow; common comfrey, purple-headed, and tiny blue speed-well, their first freshness gone. Groundsel, however, has many seeds and tall dock lifts seeded spikes. One Spring-like flower of chervil blooms late near orange berries of wild arum. A rare clover has been found by a clouded yellow butterfly and it has settled for nectar before migrating.

Dandelion clocks fluff around on the air, carrying their seeds over a wide area. A seedless head shows among the rough grass. This was known by the Normans as a 'priest's crown'.

I miss the swift-winged martins flying above the torquoise leaves of the willows. They have now departed for South Africa. More spiders weave around this October than wasps fly, but the Queen wasp has hibernated and the community expires, while blue-bottles seek refuge indoors.

After rain, 'fairy' rings of parasol mushrooms, coffee and cream in figures of eight, curve the lawn, as they do each year. Some 'fairy rings' are reputed to be hundreds of years old. Indoors, near the hot-water tank, I find a yellow brimstone butterfly preparing to settle for the Winter.

The cold persists and the garden begins to look bedraggled, with some birds visiting for food. The scattered remains of harvest in the fields must be depleted. The only colour now that is reminiscent of Spring comes from the leaves of a few vividly orange marigolds and the greening, sun-flushed heads of hydrangeas. But, even these, soon become ragged. All greens are no longer pure, but blueish or yellow-brown.

Giant sunflowers, now shrivelled and brown, hang their heavy heads on weakened stems, their first splendour gone. They are still, however, a great favourite with birds and insects. I am cheered by the sight, almost hidden, of a violet campanula bell and a blue forget-me-not.

I am disappointed not to have seen this year, that rare butterfly, the comma, so called because of the 'comma mark' on its wings. Once found only in the Wye Valley, it has now made its home in Hertfordshire. The second brood will be about, but it will soon hibernate, hanging like a dead leaf from the sheltered side of a tree trunk.

One small rose bush has had through the year, red-flecked leaves, most unusual. Happily, dark red, pink and white roses still bloom in single beauty, a gladdening sight on a grey day, under mourning skies, with the smell of smoke heavy in the air.

The twentieth, and a great change in the weather. Roads are temporarily closed in torrential rains with high winds. An afternoon break in the clouds brings a loud twittering of wrens and other small birds from the coppice. A snail is quickly snapped up by a blackbird and my robin concentrates anxiously for centipedes and insects. Magpies are again in evidence, dropping in a splendour of black and white to feed briefly on the lawn. From a hole in the grass, a late wasp creeps drunkenly.

My toad has left his little pool and may wander two or three miles or just to the compost heap, before he hibernates, deeply hidden, only appearing on damp days to feed. I miss the missel-thrush. Expect he has forsaken the coppice to roam in flocks and feed on rowan berries of the woods. He should find plenty this year.

A buddleia near is past its prime, but the purple plumes are beloved by many colourful red admirals and large tortoiseshells, hoping to feed well before the rigours of Winter.

Bullfinches are gathering now, choosing my few remaining poppy heads on which to feast. From a low bough my keen-eyed chaffinch dives adroitly down to snap a passing seed which is carried on the breeze. Twittering, a tiny goldcrest perches, later in the day, on a fir in the coppice. He is our smallest bird and sometimes probes upside-down for spiders.

The weather is fresh with much sunshine. Silver discs of honesty are exploding, releasing their seeds. A few precious flowers linger on withering sweet pea plants, welcome in this month of dwindling flower faces. On a seeding thistle, a peacock butterfly, with lovely blue-mauve 'eyes', sits motionless.

A jay falls, screaming, into the garden, imitating forest birds, scattering all small birds. He feeds now mostly on the beech mast of the coppice, but in the Spring is a great robber of eggs and baby birds. For the Winter, he hides nuts under tree roots and often flies far afield for food. I watch him. How beautiful he is, with his warmly pink breast, black and white tail and the wondrous blue in his wings.

The dampness and wind passes and October, with the nearness of Winter, makes her last offering of warm, golden days. My robin chirps, happily claiming his territory of the lawn, which is opposite to that of my blackbrid. Blue tits, too, are back in the garden.

Today, a little breeze stirs the tops of the sycamore tree, the 'lock and key' tree, where the leaves are as yellow this year as beech. Many of the hard, brown seeds, in 'twin keys', are loose and are spinning to the ground.

It is a radiant day, perfect for visiting my friend's orchard, where a new scent, that of ripeness, pervades the changing trees. A hedge encircles the orchard, partly as a wind-break to preserve the blossom and also because of the wild bees which nest at its base and are necessary for the pollination of the fruit.

Once a sea of blossom, chequered shadows now fall under the heavy laden branches, where leaves and fruit ripen together. The grass below is strewn with the wind's work, fallen apples, pears and plums, unripened or splitting to exude their goodness. For many years beneficial to man, this fruit is appreciated by others.

To this decaying fruit a hedgehog comes at dusk, leaving his nest in the dense undergrowth, for the slugs which are often present. Butterflies, wasps and bees all love the syrup of fallen fruit and crab apples.

The goldfinches that have bred here also like the insects that hibernate in apple trees, but they seem to love thistles even more and once were called 'thistle-creepers'.

Under an almond tree, the ground is covered with shells, where a squirrel has been busy hiding his nuts in several places as a Winter store. In a branch of the pear, a dormouse has thriftily built his round home with a side entrance and where fruit will be at hand and plentiful moths are hidden. This Spring he had six babies, like tiny, pink balls. When grown a little, they are graceful in play and move from place to place, tail to long tail, like a schoolgirl crocodile. Among the orange and apricot leaves of a cherry tree, a coal-tit feasts.

What a wonderful tree is shown this year, by the Royal Horticultural Society. As small as three inches, it yet bears sixty ripe and rosy apples. This month, I find another noble tree, a cedar, whose purple-black branches spread wide-lifting wings. It must have stood so for centuries. It was John Evelyn who first introduced these trees to our country. Beneath the cedar, I discover another treasure of October, goblets of the Autumn crocus, richly coloured in lilac, rose and ruby. The shining leaves of Spring have died and now the delicate flowers are rising like the phoenix from its ashes, out of the decayed Autumn soil. The 'crocus' is actually a lily that brings an added beauty to this month.

On Merry Hill, twilight falls and evening shadows steal. Trees are silhouetted against skies of russet and lavender. Over the fields rises a thin veil of mist. The swallows have gone, leaving an emptiness that cannot be filled.

'Dusky it grows, the dews descend,
The night wind stirs the fern.'

Birds are roosting and the world is quiet. One lonely and solitary leaf falls softly, silently. An owl, hunting, calls from the Hill, his ghostly, white shape floating, disembodied, unreal, through the gloom. A gleam of his strange, yellow eyes and he is gone.

A bloom hangs in the night over the darkness of the garden. The year's last roses glimmer, giving a sweet scent.

'Now sleeps the crimson petal,
Now the white.'

Rich and fragrant, the air is heavy with odours and drenched with dew. I pause, while moths flit by on soundless wings.

A sudden moon and the fields seem covered in snow, with every pathway a river of milk and every flower alone in light. And I understood Emerson: 'The man who has seen the moon break out of the clouds, has been present at the creation of light.'

Its beams fall for a while in milky bronze on the little garden pool. Soon, it is a harvest moon, round and triumphant and golden, slung in moon-splashed heavens that are flooded with light.

'Full moon, not pensive, pale,
In misty veil',

bringing unreality to a world that is already breathless, still.

Apples of Silver

All morning, here, the apples hang
In discs of red-stained splendour:

Now high the moon the heavens roves
Crystal, aloof, remotely close,
Dyeing their rose to fragile hue—
Drinking their silver as the dew;
And leaves of bronze in moonlight lie
Shadowed and turquoise turned;
& paled & pearled the fruit stays still,
Unreal and motionless, untouchable
And silver binds each tawny globe,
And silver barrs the morning gold.

Clear-etched, ethereal, night's cold fruit
On burdened branches leaning low,
Loses its gold in a silvered glow.

NOVEMBER

'There is a wind where the rose was
Cold rain where sweet grass was
And clouds like sheep
Stream o'er the deep
Grey skies where the lark was.'

Since the year is so soon to be lost, the pageantry of November, though paled with subdued tones, is precious indeed. The countryside, that has known fullness of life, now grows calm, resigned, content to wait for storms that are inevitable.

It is impossible to be indifferent to Autumn. Although Charles Dickens thought it 'a time of visible decay', Edward Thomas was surely more perceptive when he declared, 'the dead, the waste, and all to sweetness turned.'

A chill invades the air, the air that quivers with nostalgic scents from smoking bonfires and warming home fires, richly redolent of the sweetness of burning apple and beech wood.

This November, Autumn is lingering. Early morning mists of the first week persist often until midday. Some of the season's splendour is still with us, as though the sun god of the pagans has gathered the flames from innumerable fires to brighten the passing of Autumn.

On Merry Hill, November's joys, though not at first apparent, are rewarding. A subtle change has come in colour and atmosphere; a darkening, a smell of damp soil, a thickening of grasses, a thinning of branches and a secretive, hidden urgency of movement among birds and small animals. All fostering the belief that life is ending. Even the sun seems reluctant to claim his ascendancy.

Guy Fawkes Day passes, with its more ancient celebration of the Autumn fire festival and, suddenly, after the mist, a delicate beauty is disclosed. And, like a drawn curtain, they resolve and the day glows, showing a hazy loveliness as in a painting by Monet.

Trees of the lane gain a new grace with solitary,

shining leaves hanging from wet boughs or flung before the breeze like 'pestilence stricken multitudes'. Berries have a startling brilliance; glistening brambles are festooned with webs of spiders.

Watching the final mists fade from the meadow, I stand by the five-barred gate. The sky is colourless, the stubble and root crops of the fields are scored with whiteness. Slowly, the cold sun gleams and everything becomes distinct. The beech is aglow and ripe with colour; red maple leaves slowly and singly fall; trails of ivy shine and holly leaves glister as though polished. Almost bare, the horse-chestnut stands firmly, its feet covered with an orange and yellow carpet. Mosses spread over the roots of the oak tree, where beetles eat the bark and the last leaves, brown and sere, cling to the branch. Below, acorns have fallen in the roughened grass.

The first Sunday and still beautiful weather. Sheltered by the wall, Cornish lily flowers are un-blemished and crisply pink, like crumpled crêpe paper. Michaelmas daisies still bloom. I linger on the lawn to watch a young magpie hesitantly leave his coppice nest to investigate alone. He is not yet as handsome as his long-tailed, black and white father, but has a russet breast.

A rotting apple is the centre of a sudden squabble between my thrush and an angry, grey squirrel. The latter, who can smell rotting fruit from the coppice which is a hundred yards away, wins and departs in triumph with his prize. Not content, he joins his family in stripping bark from trees.

My disconsolate thrush has to be satisfied with a juicy worm. These birds, though seemingly shy, can be tamed somewhat. My friend was delighted when, after

many attempts, her thrush accepted cheese from her hand.

Under a rose bush, on slim, drooping stems, a few delicate, pink fuchsia flowers still hang. The magnolia tree has gained a new beauty in the thick leaves, which are now mellow and saffron. Some have fallen, disclosing next year's buds, which are tight and surprisingly large. Ice plant is burnished and, here and there, a mauve candytuft shows, while stinging nettles grow happily on the compost heap and a solitary marigold makes a blaze of colour among all the dull greens.

Under skeleton cabbage leaves, my chaffinch is probing for caterpillars, while a large white butterfly rises from its second brood in a flapping flight. Only a few red-tinted leaves are left on the sumach tree, but its bare branches have their own grace. This tree was introduced into England in the fifteenth century and its empty outline is windswept and still intriguingly Oriental.

The second week brings a pale gold sky and there is a heaviness everywhere that augurs change. From my sycamore, a lonely disc of yellow flutters down. But, except for one desultory, small bird that flies from the coppice, there seems to be no movement. The air is deeply still, with a dense silence.

As I watch, a green woodpecker suddenly falls on the lawn for worms and yellow ants. This bird, with his green-yellow feathers and red cap, was once sacred as a 'rain-bird' to Thor, god of thunder. He is soon gone and my hedgehog reappears, moving unusually quickly and gobbling small grubs as he goes. He is a strange animal for he swims and even climbs trees for birds' eggs. Unafraid of the adder, he nips him in the neck before eating him. Only the fox outwits the hedgehog, luring him into the water and then attacking his exposed head.

Mid-month and a day of rain, followed by fog and early darkness at 4 p.m. Has November, briefly deferred, finally arrived? The fog blankets the fields and, in the road, it whirls around, yellow and sulphurous, seeming to rest in concentrated clouds on each lamp, making its light hazy and mellow.

More rain and unkempt starlings crowd the lawn. Worms and centipedes on the surface are rapidly snapped up by my blackbird. Under the eaves, I find a wasp crawling heavily around and getting nowhere, his wings forgotten with the suns of Summer.

A cold, clear sky and parts·of the garden are decayed and sad, its scents gone with the colour. Above the coppice curls the smoke of a bonfire. The harsh cries of magpies are no longer heard and squirrels have deserted us, busy in their search for nuts and the restrengthening of their dreys.

A sudden shower of sleet and most plants show bedraggled stems with only a few damp blooms. Chrysanthemums and dahlias are still richly coloured, but hang, heavy-headed, while lone roses, white and pink, brave the indecisive weather.

Merry Hill is damp and deserted. I linger to watch, as a baby squirrel that has fallen into the road from an overhanging tree, is rescued and returned. A wild, west wind sweeps across the fields, stripping clear and bare the trees' boughs. Many rooks are gathering. My meadow has undertones of grey and in its rough grass no lark sings.

Shepherd's purse is fading, with tiny heart's-ease, that plant which is still given to racing pigeons as a medicine. Across the meadow, on an empty elm bough, a fieldfare perches. Like the year the tree, too, is dying. Wood pigeons come for food in the fields where they often decimate the crops. More gusts of wind from cold skies and layers of leaves are driven, 'like ghosts from an enchanter fleeing'.

Today, November is quite chill and dark with gloomy bushes and bereft trees. Yet, slowly, the sun breaks through the faint mists, softening every contour of the rounded hills. Although most of the 'keys' have fallen, my sycamore boughs still cling to their little wings of pale gold leaves. The missel-thrush has returned, if only briefly, to peck among the remaining seeds. The sun gives little warmth and that reluctantly. In the hedgerows, above seeded dock and cow parsley, are brilliant berries of the rose, rich in vitamin C. A finch probes with his strong beak at the hawthorns' fruit. Blackberries have ripened and are depleted, honeysuckle has clusters of orange, and purple-black fruits show on the deadly nightshade. Feathery parasol seeds of dandelion and traveller's joy float on the air, but the ivy's pale green petals are gone. From the smell of bonfires and the heavy scents of Autumn I turn homewards to savour the delicious odour of an apple-

wood fire.

Two unexpectedly warm days follow and my Cornish lilies glow, untouched by wind and rain, their flowers on long, leafless stems still delicate and fine. But Winter is coming and the grass snake and his family have long since hibernated in a cosy hole of the lane's bank. The stoat is getting hungry. His long white body, low to the earth, slinks along unseen by the small animals he follows. I watch, fascinated, as a stag beetle burrows for his home and an enterprising chaffinch shakes the stem of a burdock and eats the falling seeds.

More rain and the sodden fields mourn. Yet the earth is soft with a freshened colour. The cold intensifies into showers of sleet. It is too wet now for the plough and rain beetles, cockroaches, mice and rats are driven to warmer hiding places in outhouses and barns. Chill winds rob most of the deciduous trees of their last, lingering leaves.

The third week and the day brings a special joy, the quiet beauty of a few blooms of herb Robert. Through the withered vegetation of the hedgerow, the shy, pink flowers show, stained with red, the hairs on their stems silvering in a rare ray of sunlight.

There are shades of blackness, mysterious and deep, in Whippendell Woods. It is bright and cold and a delight to walk, crisping the dry, crackling carpet of leaves. Fallen, they make richness for their parent tree. Under its canopy, the needles soften the path beneath my feet.

Disraeli said, 'A forest is like an ocean.' It certainly has moods and moving currents and waves of glinting light that disclose the tinted beauty of bark, and gild and bronze the layered leaves of the ground. To savour the quality of a wood, one must be alone; alone with its special silence and its green, translucent light; alone with the mystery of trees and the quietness of life above, around and beneath. Like an old church, an ancient wood seems to hold within it the gathered thought and love of centuries. For

'There's not a leaf that falls upon the ground
Buts holds some joy of silence and of sound.'

Where beeches lift, the woods are roofed and floored in gold. In avenues of meeting boughs, smooth, grey trunks hold branches like hands, beckoning skywards. I touch one rounded, marble-like column and find a small fungus has grown with creamy heads.

There is a fine crop of mushrooms and toadstools and this is the month when their colour deepens. They spring from fallen leaves and, in their dying, help to feed the tree. Toads were once believed to be poisonous and toadstools grow in the same dark, damp places that toads frequent. Toadstools were also thought to be the 'stools' of fairies. Certainly, to imaginative children, both mushrooms and toadstools have always held a magical quality, perhaps because they seem to appear magically, overnight, from nowhere.

I pass a gnarled old yew tree which is partly covered with a bracket fungus, by a cedar which holds out plumed branches. Wild cherries grow, the rowan fruits are red and on the crab apple tree globular fruits are ripening. Ancient hollies, thick-stemmed, grow here, yellow-berried and tall as young larch. On the wood's margin I find the deathcap fungus, green-yellow and poisonous. Here too, are the orange-red fingers of the caterpillar fungus, in which a dead caterpillar is embedded. The most unpleasant smell of the stinkhorn fungus meets me. Swarms of flies, attracted by it, cover this fungus, spreading the spores as they fly away. In the grass of the verge, yellowish puffballs show, which wait for a raindrop to make them explode and release their many spores.

On a miniature hill, which may be an ancient burial ground, three mighty pines stand, with wine-red trunks, immensely tall and straight. Underneath, are chanterelle toadstools, flower-like, tunnel-shaped and yellow as egg yolk. Near by, is the brilliant orange, white-spotted fly agaric, growing under a birch tree. Like colourful umbrellas, they are beautiful but poisonous. Once, however, they were eaten by Viking warriors to stimulate them for battle.

The Romans were fond of fungi, but they were eaten before them, by prehistoric man. Attractive spore patterns can be taken of these mushrooms and toadstools by laying the ripe cup on white card and removing carefully.

In the delicate outline of leafless trees, November reveals new beauties. I wander between the trees, smoothing their bark. How varied is the bark, in shape, colour and texture. One day I will return to make bark rubbings of the many trees of the wood. Dogwood is enhanced by moss. From 'the forest's ferny floor', graceful birch are slender and silver. The vast sweet chestnut, that has stood for innumerable years, is solid and secure in the earth, with its great-girthed trunk whorled around in yellows, ochres and acid greens. How many animals, birds and insects it must have nurtured and sheltered. Titmouse, woodlice, nuthatch, owl, the list is endless.

Standing in the heart of the wood, I listen to its sounds. In the larches above my head, a squirrel scolds a wood pigeon with a 'Quark! Quark!' The spiders that lurk and hunt among the withering leaves are silent, so are the woodmice among the acorns. Not so, the blue-winged jay. I hear his flapping flight and his shrill voice of mimicry. He even mobs the tawny owl of the hollow tree, in his craving for sweet chestnuts, burying them, squirrel-wise, under the trees' roots.

With a whirring of wings, a cock pheasant rises and from the depths of the wood comes the rhythmic tapping of a great spotted woodpecker as he probes the bark of an oak tree, for insects, grubs and ants. With his black and white plumage and red patch, he is a fine sight. In the tree's trunk he bored for his babies a nest, in the Summer.

In a few isolated places rose bay willow-herb

persists, but fading foxgloves have taken the place of the vanished bluebells. Here and there, withered wood angelica shows among the browning bracken and gilded ferns, where glossy, splitting chestnuts lie.

On the edge of the wood, woody nightshade holds on the same stem, purple flowers and poisonous red berries, while trails of white bryony have clusters of three, orange globes. Old man's beard festoons the brambles with a veil of grey lace and, in a little swampy clearing, a snipe is probing with his bill into the wet mud of a ditch. He is searching for worms or snails and grubs and, like the hippopotamus, loves mud. From him, comes the phrase 'guttersnipe'.

There is no sign of the rarely seen dormouse, who has enjoyed his Spring and Summer climbing trees and is now eating large quantities of hazel nuts and berries. Soon, now, when he has absorbed all he can find, he will roll himself up into a tight ball and hide away for the Winter.

Another day and the sun is hidden.

'The woods are still. No breath of air
Stirs in leaf or brake.'

But, suddenly, from a darkening and bruised sky, a chill wind begins to whistle through the trees, moving the dry and dying leaves. A mist of small rain hangs across the landscape. The last of the oak leaves moves and, one by one, acorns are torn from the boughs and fall with a sharp sound among shrivelled ferns, fungus and leaf mould. Little harsh, dry sounds come from the chestnuts, beeches creak and birches tremble.

Through the wind comes the grieving note of a tawny owl where he crouches in an ivy-covered spruce and, answering, the wild cries of a roosting pheasant. The atmosphere becomes oppressive, foreboding.

Sudden, a great wind rushes through the wood; the leafy carpet is flurried in clouds and branches are tossed and beaten. A mighty onslaught of sound and of shrieking wind, and birds flee before the driven flocks of leaves.

Stillness comes. No sign of life. All is hidden, trembling in fear. The wind is gone, leaving only a moaning high in the tree-tops. And, after the moaning, 'Silence, listening to silence'. And, tentatively, through the silence, the small sound of a bird.

From a yew tree, with its lichen of vivid green, he sings, the missel-thrush, undaunted now, as though oblivious of the storm's blusterings. The last remnants of the wind threaten to shake him from his bough, but still he clings and still he sings.

And, like a spear above his head, the sun's ray pierces the black clouds and the rain draws fragrance from the fir trees. As though hurling defiance at the gale, like the notes of a melodious flute, one small bird sings. What a wonderful way to face the Winter. It is believed by some that this songster is the bird that sang three times in a storm-laden sky and was heard by the sorrowing St. Peter.

The storm has ceased and the woods resume their air of mystery, their atmosphere of the past, their assurance for the future.

The last three days and, alternating with warmless sun and rain, sleet showers fall. And, veiled in a thick fog, November is gradually fading. Yet, through the haze, dim and indistinct, the red globe of the sun appears.

This capricious month saves a surprise for its last day, when the first hard frost brings a glistening beauty. On whitened lawns and fields, footsteps crackle and there is a crispness in the air which is exhilarating.

Though the sun gives little warmth it sends sudden shafts of pale light on a magpie's rich plumage, it lightens the squirrel's grey coat, it deepens the orange of my blackbird's bill, gives radiance to the fallen leaves and makes more brilliant the breast of my robin. And, round the brightness of berries, it turns to spun silver the spider's threads. All small things, but so precious in November.

The Silent Woods

Coins gold and –
green doubloons
Fleck the waters and
fringe the pools,
And the sentinel pines
& the lime trees' spread
Form a layered trellis
of boughs overhead;
And the still woods wait
With their carpets of gold
To silently tell
Of mysteries old;

And under the wealth
Of copper and green,
Only a dog, lonely
And loping, is seen;
And glossy the holly
In greens and jade,
And glistening & green
Is the ivy's shade.

And the young ferns fade
and the bracken dies,
And the wine-red pines pierce
the vault of the skies;

All bird-song has ceased
And all squirrels sleep,
In the waiting woods,
So silent, so deep.

DECEMBER

'Happy is the man whose year has been excellent,
who looks before with hope and back with content.'
'In my end is the beginning.'

December — and Nature seems to have abandoned her quest, with the year dying in desolation. But Winter is not dead; most of it is not even sleeping. It is a gaining in strength for a new Spring. The soil is being broken down by frost and millions of microbes, bulbs, corms, roots and tubers are building sustenance for a fresh blooming. Though the seething life of the earth is buried, half-hidden and only faintly heard, the mysterious cycle of life is being forever renewed, for Winter is probably more deeply alive than any other season.

Spring's pageant has been magical, tumultuous; Summer's overwhelming, prodigal; Autumn's a riot of colour, an abundance of richness, as though to say 'Before I go, I give all.'

But what of Winter? De la Mare regrets,

'Alas, that time should be
So rich, yet fugitive, a pageantry!'

Someone else said, 'It is good to day-dream in December.' But Winter brings more than memories and the month has a certain, austere splendour of its own. Her music has no insistence, no clamour. It is in a minor key. She does not force your attention or demand acknowledgement of her gifts. Yet, for the searching eye, the discerning mind, the waiting ear, she has rewards enough. Nature still continues to amaze and delight and bring us to thoughtful wonder.

Her truly dreary days are mostly few. When mists blot out colour, form and substance, through the swirling comes the lustre of a lone flower, the brilliance of berries, the wonder of a frail web. December has grace, too, in the slow pirouetting of a gold and falling leaf, the delicate tracery of boughs, the flash of a jay's wings. New-found treasures are the magical green of the first crocus tip, or a robin's breast, when it turns to pure flame in a sunset glow. And only in Winter can one wonder at the perfect designs of snowflakes or listen to the mighty music of the wind in the trees.

The first day of this December is bleak and melancholic. From my window, life in the garden seems suspended and the coppice deserted. Everything looks dead, broken, bowed, sterile and shrivelled and snow leadens the skies. I remember Robert Browning,

'So the year's done with,
All March begun with,
April's endeavour;
May wreaths that bound me,
June needs must sever;
Now snow falls round me.'

For consolation, I seek my garden and two things never disappoint. Large buds show clearly on my magnolia and some rose bushes hold a few late buds. Some of them will flower, even if most will be frosted. I gather a spray of hardy, yellow chrysanthemums and one dark, red dahlia, with a marigold to brighten and a sheltered snapdragon to cheer. These, indoors, with silver honesty, are a decorative delight, for, in memory,

'The old splendour
And magical scents, to a wondering eye, bring
The same glory.'

Afternoon and a luminous light gilds the lichen on the apple tree. No chaffinch; he has gone away to roam in flocks in search of seeds and beech mast. Wood pigeons invade the coppice, somehow evading truculent magpies and the occasional, foraging squirrel. Two ring-doves, with white bands around their necks, visit the lawn on laboured wings. They have a mournful cry.

In medieval times, doves were much cultivated, a dove-cote being a sign of manorial rights. Vast, circular, round-roofed and sometimes made of flint, they housed hundreds of birds. The soft blue and pink plumage of my visitors is warm, but they are careless creatures. Their nests are so clumsily built that often the eggs fall out. It is the only bird, however, to produce a kind of milk with which to feed its young.

Recently, one pigeon pecked so fiercely outside a window that was being cleaned, that the pane actually broke. Did he imagine that his reflection was another bird that was taking his precious food?

In order to save it from extinction, the pink pigeon is now being bred. Sometimes a dove is used as a foster parent, since the careless parent often tramples her egg.

How amazing are homing pigeons. In wartime their inborn instinct saved many lives. Now, they are being used to raise money for charity. It is believed that the sun helps them to navigate in their long flights of over five hundred miles. Flying at speeds of sixty miles an hour they have, like other birds, remarkable stamina.

Barely begun, December gives another still, bright day and two yellow roses bloom gloriously and I pluck Christmas roses with frilly, dark leaves. White and pink, they are a small treasure, their pearly petals easily bruised, as delicate as the windflowers of the Spring. One tall-stemmed, Japanese anemone is out and dainty Winter jasmine, sweet-smelling. At my window, a few heart-shaped leaves of clematis glance, like amber glass, pale and translucent. All coppice trees are bare now, except the Scotch pine and, in the avenue, the willow alone is loath to part with its silvered leaves.

It is late afternoon and many birds are active in the conifers. A bullfinch searches in the apple tree and blue tits are busy among the berberis berries. Wrens and sparrows join them for titbits on the lawn. This month they eat with serious intent and fly with purpose. A great spotted woodpecker swoops and all small birds flee. After tapping for beetles in the bark of trees, this finely coloured and rare visitor turns its attention to the insects of the damp grass.

Through the stillness and the icy air, a noisy clatter from the coppice. Two squirrels and a magpie are quarrelling raucously, the latter magical in flight with slender wingtips.

No movement, no sound. The sun goes down like a blood-red orange and the skies are flooded with lemon, gold and salmon, a sky to linger with. One of the glories of December.

The cold deepens through the next few days, with fog, followed by rain. For once, Merry Hill is dark and uninviting, its verges rutted with damp, brown leaves. Under a lowering sky the scene is drear and the old barn, where children gather straw for their rabbits, is becoming depleted. With the barrenness of fields and the stark tangle of hedgerow, the loneliness is emphasised by the monotonous cries of circling rooks.

All merriment and life seems gone from the Hill, with the earth naked and crusted. No flowers are visible, but my robin goes before me as always, curiously flitting from branch to branch. And I am reminded of Emily Brontë:

'What matters it, if but within
We hold a bright, unsullied sky
Of suns that know no Winter days?'

High above me, in the coppice, where a bushy tail moves rapidly, a squirrel's drey is covered with thorny twigs. There is a decidedly pointed tip to the tiny black ash buds and those on the beech hold a hint of bronze. The biggest buds on the chestnut tree already betray a slight stickiness and long cones hang from the spruce, holding inside their precious seeds.

Straggling briers still cling to an odd leaf of brilliance and the faint, watery sun lends a soft light to trails of old man's beard. A fieldfare, hopeful for food, pecks near the shining berries of holly. Seemingly empty, the brambles hide the robin's nest and crisping ferns conceal tiny insects.

Venturing from retreat in an old bird's nest, a wood mouse noses around for a little food. He loves hips, buds, acorns and snails and even raids a bees' hive for honey.

The second week and snow falls thickly on Merry Hill, making it a landscape enveloped in a white sheet.

'O, tranquil, silent snow,
Such loveliness to see!'

In the afternoon, a suggestion of sun breaks through and a golden sunset makes beautiful the black, writhing arms of the doomed elms on the far hill.

Darkness falls. A moon, almost round, gleams on the dark trace of twig and bough, lighting with jewels the encrusted snow.

A new morning and ice covers the deep snows. Over the fields, a bitter wind blows drifts in horizontal gusts, blinding and obliterating. On the lawn, clear footprints of animals and birds cross and recross, in an apparent search for food and warmth under the eaves. Squirrel, rabbit, fox, cat and large birds have left their mark on the hard, frosted lawn.

The next day and all day, a grey fog persists. Man and beast shelter. Through the gloom, a clump of cabbages emerges, iced with frost patterns and bedraggled and tattered chrysanthemums rise bravely. Like flimsy sleeves, the mist streams through open doors. Later, a little sun, in patches of cold blue sky tries, briefly, to show, but, in a thick and blinding curtain, three more inches of snow falls, stifling all sound in a glistening world.

More heavy blizzards and the half-month ends with temperatures well below zero, and the coldest night for a century.

Gales and snowstorms are severe. The village pond and the Pools freeze. Skating is enjoyed in the frozen Fens and Merry Hill rings with the cries of sliding and tobogganning children. A puffin lands in the streets of Leighton Buzzard, blown miles off his course. In Devon lightning is severe. On the river Dee, ice floes sweep down to crush and sink small fishing boats. In one Wildfowl Reserve, black-throated Siberian thrushes which come from Russia, must have been vastly disappointed and Arctic geese are flying away from the abnormal, sub-zero temperature.

Schoolgirls in Yorkshire are pulled to school in sleighs by deer and the wide lakes of Windermere and Coniston freeze. In an attempt to browse from trees, deer are standing on their hind legs and keepers are having to break down boughs of leaves for them. As

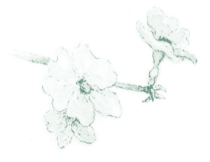

there is so little grass left and that buried deep in snow, the deer nibble at the bark of trees. But, surviving the rigours of Winter, they can live for twenty-five years.

While we shiver or delight in the cold, frost and snow, for rich Arabs and their camels, it is the time of hawking in the desert. Here, their trained falcons hunt the desert hare and the houbara, a rare bustard, which breeds on the Russian steppes and goes to the Arabian desert for the Winter. To the dismay of conservationists, many falcons are lost and the houbara is becoming still more rare.

Wildlife does not fare as well as man. Squirrels and jays cannot get at their stores of nuts. Worms are safely hidden from thrushes and many more small creatures. As the sparse grass is hardened with frost or covered with snow, farmers have to feed and rescue sheep and cattle buried in drifts. This Winter many unlucky sheep are marooned on islands created by floods from swollen rivers. Ironically, the tinkling sound comes not from their bells, but from the icicles on their fleece.

Evening and a frosty sky, with early stars showing and an incredible sunset in flames and icy greens. A tom-cat howls on the roof top, claiming his territory. Recently, one was trapped in a chest of drawers for three weeks, but emerged unscathed, having lived on his own body food.

Mid month and the last of the willow leaves hang in icicles, quite frozen. Rain and sleet blizzards dispatch most of the snow. One golden rod has survived the wintry onslaught. Small birds hop near for scraps and water. Over white drifts of the lawn, long, mauve shadows fall from the trees of the coppice, where the magpie looks splendid with his black and white plumage and trailing tail against boughs of black. He is busy dislodging the clinging snow from his lofty, untidy and domed nest, which now is plainly visible. He needs it as a refuge and, later, for his family.

Over the forsythia and elder bushes hangs a haze of wine-colour. I gather with care, one golden rose-bud, somewhat blighted, a spray of purple heather and a pink lily flower which has survived. My dwarf golden cypress, outlined in crisp snow, looks like a walking penguin, leaning forward over the lawn.

Another snow blizzard on the twentieth. Thick, thick snow and a little rain. Night's frost and fog makes it treacherous underfoot, where fine, powdery snow has covered yet again a film of ice. Alas, my willow is now quite bare, icicles having torn away the solitary leaves in a slight thaw.

Flurries of snow for two more days, but the flakes melt as they fall. Yet the bitter cold and ice are forever

with us. In the coppice magpie and rook eye each other defiantly, while, from a safe distance, smaller birds watch. My blackbird is the first to emerge for food. His appetite is insatiable. Hunger drives the thrush more quickly than usual to the fore. In the sparkling air, he seems more brightly bespeckled than ever.

The blue tits are most enterprising, acrobatic and endearing, swinging upside-down on the nut-bag string, pecking milk tops for the cream beneath and actually selecting the gold top, if there is a choice. One family of blue tits has discovered the warmth from light bulbs in porches and perch happily close to the lit bulb, each bird returning nightly to his own bulb.

Christmas Day and the morning dark and whitely shrouded. The cold is intense. Mists over the fields of Merry Hill and wreathing the tree-tops, but the frost transforms the brambles into strings of glistening jewels. Yellow-billed, my blackbird goes through his ritual of cleaning and preening, until each feather is shining and lubricated.

Three days left and a hint of thaw. Flocks of gulls but no small birds. Several rose-buds show with frozen petals, from which moisture slowly oozes. Still,

'When the rose is faded,
Memory may still dwell on
Her beauty shadowed,
And the sweet smell gone.'

The ending of the year and sudden rain. A pale, blue sky and my thrush, robin and wren all revel in the water of the hollowed lawn. Although the willow is bereft of its leaves, the holly gleams brighter in my little avenue, where a blackbird sings his welcome and beloved song as I pass.

This year of 1981 slips away in white fogs, rain and thaw. The last day is wreathed in thick, swirling mists. Heaviness hangs on the air like a blanket. On Merry Hill, even when the fog withdraws, footsteps are hushed, birds cease to sing and the noise of distant traffic is blurred.

Then, as though to assert his ascendancy, a late sun, in sudden energy, shines brilliantly in the far west. In the sky, long low bars of silvered grey and, reaching upwards, gold-edged clouds of delicate rose and blue and green. Behind the pine, a few faint stars appear as, serene with 'the silent music of tranquillity', night falls.

And, beyond the coppice,

'With far-flung arms enflamed with light.
The trees are lost in dream.'

The Wild Swan

—Silent and still, in the reeds' shadow deep,
Curved head down-folding, the swan lies asleep:
Till from her dreaming, the white plumage stirs,
Trembling anew, neath the high-lifting firs;
Quivers and wonders, all milk-white and cool,
Mirrored and shimmering, down by the pool;

—Slowly the sleek head no longer is bowed,
Smoothly, the slim neck is uncoiled and proud;
Snowy and silk-soft, it lifts high and long,
Unfurls each wide wing, swift, still and strong;
Powerfully poised, the white pinions lift,
Stream lined & slender the wild swan's adrift.

Up and away, like a pale moving dream,
Trailing & lost, in the cloud's endless stream.